SpringerBriefs in Applied Sciences and Technology

For further volumes:
http://www.springer.com/series/8884

Doreen E. Kalz · Jens Pfafferott

Thermal Comfort and Energy-Efficient Cooling of Nonresidential Buildings

Doreen E. Kalz
Fraunhofer Institute for Solar Energy
  Systems ISE
Freiburg
Germany

Jens Pfafferott
Offenburg University of Applied Sciences
Offenburg
Germany

ISSN 2191-530X   ISSN 2191-5318 (electronic)
ISBN 978-3-319-04581-8   ISBN 978-3-319-04582-5 (eBook)
DOI 10.1007/978-3-319-04582-5
Springer Cham Heidelberg New York Dordrecht London

Library of Congress Control Number: 2014931763

Printed on acid-free paper

Springer is part of Springer Science+Business Media (www.springer.com)

Preface

This guidebook on low-energy cooling and thermal comfort supports HVAC planners in reducing the cooling-energy demand, improving the indoor environment, and designing more cost-effective building concepts.

High-performance buildings have shown that it is possible to go clearly beyond the energy requirements of existing legislation and obtaining good thermal comfort—both in summer and winter. However, there is still a strong uncertainty in day-to-day practice due to the lack of legislative regulations for mixed-mode buildings under summer conditions—buildings which are neither only naturally ventilated nor fully air-conditioned, but use a mix of different low-energy cooling techniques.

Most of the new nonresidential buildings are mechanically cooled with low-energy techniques using a mix of ambient heat-sinks such as ground, night, or evaporative cooling. Moreover, more and more retrofit projects have also been using mixed-mode low-energy cooling in recent years.

Based on the findings from monitoring campaigns (long-term measurements in combination with field studies on thermal comfort), simulation studies, and a comprehensive review on existing standards and guidelines, this guidebook gives a pathway toward a successful implementation of passive and low-energy cooling techniques in energy-efficient nonresidential buildings.

Acknowledgments

Some of the material presented in this publication has been collected and developed within the Project THERMCO—Thermal comfort in buildings with low-energy cooling; establishing an annex for EPBD-related CEN standards for buildings with high-energy efficiency and a good indoor environment, EIE/07/026/SI2.466692. The guidebook is a result of the joint effort of nine European countries. All of them contributed by collecting information and carrying out a long-term monitoring campaign in nonresidential buildings.

Furthermore, the meta-analysis of European buildings is extended by monitoring results from German nonresidential buildings. Monitoring and evaluation were funded by the Federal Ministry of Economics and Energy (BMWi) under the programs "Energy-Optimized Building" (BMWi 0335007P/C), "Energy-optimized construction in refurbishment" (BMWi 0335007C), "LowEx:Monitor" (BMWi 0327466B), and "ModQS" (BMWi 0327893A), as well as by the Federal Ministry of Transport, Building and Urban Development (Zukunft Bau), which is gratefully acknowledged.

All the people who have contributed to the project are gratefully acknowledged: A. Wagner (Karlsruhe Institute of Technology, Germany), B. Olesen and P. Strøm-Tejsen (Technical University of Denmark), J. Kurnitski (Helsinki University of Technology, Finland), M. Santamouris and T. Karlessi (National and Kapodistrian University of Athens, Greece), F. Allard and Chr. Inard (University of La Rochelle, France), K. Kabele and M. Kabrhel (Czech Technical University), L. Pagliano and P. Zangheri (Politecnico di Milano, Italy), A.-G. Ghiaus (Technical University of Civil Engineering Bucharest, Romania), O. Seppanen (REHVA, Belgium), W. Warmuth and J. Farian (PSE, Germany), M. Sonntag (Fraunhofer ISE, Freiburg), as well as G. Vogt, and F. Hölzenbein.

Monitoring, data acquisition, and commissioning are challenging tasks. The authors would sincerely like to thank the various evaluation teams for their excellent support, discussion, and cooperation: I. Repke, P. Obert, G. Mengedoht, G. Lindemann, M. Ehlers, F. Ghazai, C. Sasse, D. Schmidt, J. Kaiser, M. Kappert, C. Prechtl, R. Koenigsdorff, S. Heinrich, T. Häusler, E. Bollin, M. Melcher, T. Knapp, and B. Bagherian.

A range of international experts provided input and commented on the underlying methods on monitoring, questionnaires, and data evaluation. Their comments and suggestions were of great value: Fergus Nicol, Richard de Dear, and Sebastian Herkel.

Contents

Abbreviations

AA	Ambient Air
AC	Air-Conditioning
ACH	Air Change Rate
AHU	Air Handling Unit
AMC	Air-Based Mechanical Cooling
AVG	Average
BHEX	Borehole Heat Exchanger
CH	Chiller
CHP	Combined Heat and Power
COP	Coefficient of Performance
CP-w	Water-Based, Ceiling Suspended Cooling Panel
CT	Cooling Tower
DC	District Cooling
DH	District Heating
E	Electricity
e	Exterior
f	Free
Fin	Final
HP	Heat Pump
HR	Heat Recovery
HX	Heat Exchanger
i	Interior
m	Mechanical
Max	Maximum
Min	Minimum
MMC	Mixed-Mode Cooling
MV	Mechanical Ventilation
n/k	Not Known
NV	Night-Ventilation
PC	Passive Cooling
PMV	Predicted Mean Vote

POE	Post-Occupancy Evaluation
PPD	Predicted Percentage Dissatisfied
Prim	Primary
SBS	Sick Building Syndrome
SPF	Seasonal Performance Factor
SU	Split Unit
T	Temperature
TABS	Thermo-Active Building Systems
Therm	Thermal
W	Week
WMC	Water-Based Mechanical Cooling
Y	Year

Nomenclature

$\theta_{o,c}$	operative room temperature
θ_e	actual ambient air temperature
$\theta_{e,d}$	daily mean ambient air temperature
$\theta_{e,d,max}$	daily maximum ambient air temperature
$\theta_{e,month}$	monthly mean ambient air temperature
θ_{rm}	daily running mean ambient air temperature

How to Read This Guidebook

This guidebook is based on the evaluation of realized low-energy buildings all over Europe, using a standardized method based on existing monitoring data.

1. *Impact of Cooling on Energy Use*
   First, we take a look at the current and future cold market in Europe and define low-energy cooling concepts for nonresidential buildings. Low-energy buildings have to meet various building-physical requirements.

2. *Thermal Indoor Environment and Guidelines for Thermal Comfort*
   A brief introduction to various aspects of thermal comfort in air-conditioned, passively cooled, and mixed-mode buildings.

3. *User Satisfaction with Thermal Comfort in Office Buildings*
   Users feel satisfied with thermal comfort in the monitored low-energy office buildings. Due to the different expectations on interior thermal comfort, it should be evaluated in passively cooled buildings as well as in buildings with free and mechanical night ventilation in accordance with the adaptive-comfort model in EN 15251. By contrast, the thermal comfort in buildings should be evaluated according to the PMV model, when air-conditioning systems, fan-coil cooling, or water-based radiant cooling systems are employed. Even in these buildings, users adapt to the prevailing outdoor climate conditions. However, they only tolerate slightly higher room temperatures as compared to the temperature setpoints of 24.5 °C ±1.5 K as defined in EN 15251.

4. *Methodology for the Evaluation of Thermal Comfort in Nonresidential Buildings*
   There is a strong uncertainty in day-to-day practice due to the lack of legislative regulations for mixed-mode buildings, which are neither only naturally ventilated nor fully air-conditioned but use a mix of different low-energy cooling techniques. The practitioner receives a practical approach on how to evaluate thermal comfort and energy use in nonresidential buildings with low-energy cooling.

5. *Thermal Comfort Evaluation of Office Buildings in Europe*
   Case studies in eight European countries provide precise information concerning good- and best-practice examples. All buildings were evaluated by

using the same approach. It is clearly demonstrated that it is feasible and valuable to compare different cooling strategies based on a consistent methodology. Furthermore, this methodology can be applied in day-to-day practice in the planning, commissioning, and operation of buildings since these parameters are well-established and easily accessible.

6. *Application of Cooling Concepts to European Nonresidential Buildings*
   High performance buildings have shown that it is possible to clearly go beyond the energy requirements of existing legislation and to obtain good thermal comfort.

7. *Thermal Comfort and Energy-Efficient Cooling*
   Mixed-mode- and low-energy-cooling buildings provide good thermal comfort with a limited cooling capacity, e.g., enabling ground-cooling in combination with thermo-active building systems or part-time active air-cooling. A comprehensive assessment procedure considers thermal comfort, the energy demand for cooling, and overall energy consumption.

This guidebook provides design guidelines for architects and HVAC-engineers working on typical building concepts in the European climate zones.

Chapter 1
Impact of Cooling on Energy Use

Abstract Cooling of the built environment is a relatively new and rapidly expanding market in Europe. The pressures associated with energy efficiency call for combining energy-conservation strategies as well as for energy-efficient technologies in order to reduce a building's carbon footprint. Low-energy cooling technologies improve the users' thermal comfort and reduce the energy demand for cooling—and often also for ventilation. However, buildings need to meet minimum requirements for the application of low-energy cooling due to the limited cooling capacity.

Buildings are one of the heaviest consumers of natural resources and cause a significant portion of greenhouse gas emissions affecting climate change. In Europe, buildings account for 40–45 % of the total energy consumption, according to (EUR-Lex 2002 and EEA 2006). In the United States, greenhouse gas emissions from the construction sector have been increasing by almost 2 % per year since 1990. CO_2 emissions from residential and commercial buildings are expected to increase continuously at a rate of 1.4 % annually until the year 2025 (Brown et al. 2005). Given that buildings are responsible for approximately 20 % of the greenhouse gas emissions, there is a growing awareness for the important role buildings play in reducing their environmental effects (Stern et al. 2006). On the one hand, emissions associated with buildings and appliances are expected to grow faster than those from any other sector. On the other hand, reducing energy consumption in buildings is estimated to be the least costly way to achieve large reductions in carbon emissions (McKinsey 2007).

Pressures associated with energy efficiency call for combining energy-conservation strategies as well as for energy-efficient technologies in order to reduce a building's carbon footprint. Two key targets were set by the European Council in 2007: First, "a reduction of at least 20 % in greenhouse gases by 2020," and second, "a 20 % share of renewable energies in EU energy consumption by 2020" (COM 2008). Following Article 9 of the EPBD Recast 5/2010 (EPBD 2010), member states shall ensure that by 31 December 2020, all new buildings are nearly zero-energy buildings. Incorporating renewable energy as well as energy-efficient and sustainable design features into buildings allows for the reduction of both the resource depletion and the adverse environmental impacts of pollution generated

D. E. Kalz and J. Pfafferott, *Thermal Comfort and Energy-Efficient Cooling
of Nonresidential Buildings*, SpringerBriefs in Applied Sciences and Technology,
DOI: 10.1007/978-3-319-04582-5_1, © The Author(s) 2014

by energy production. Sustainable and energy-optimized buildings attempt to harness their architecture and physics in order to provide a high-quality interior environment with the least possible primary energy consumption.

1.1 Current and Future Cold Market in Europe

Cooling off the built environment is a relatively new and rapidly expanding market in Europe. Its growth is motivated mainly by the rising standard of living, which has made this type of equipment affordable. At the same time, people's requirements on comfort have increased. The demand for comfort cooling has been steadily increasing in all European countries, both old and new EU member states, as well as in the accession countries. Market experience shows that once 20 % of the office space in a city is air-conditioned, the rental value of noncooled spaces decreases (ECOHEATCOOL 2006). The European ECOHEATCOOL Project presents an overall definition and description of the European cooling market and its potential growth. The study concludes that the potential cooling demand and the pace of expansion for the European cooling market are greater than earlier indications. However, a development toward the cooling saturation level found in the USA (of 70 % for the residential and 73 % for the service sectors) is probably unlikely, due to differences in climatic conditions (ECOHEATCOOL 2006). Considering the current market trends in Europe, a saturation rate of 60 % for the service and 40 % for the residential sector is assumed to be realistic. This would result in a fourfold increase of the cooling market between 2000 and 2018, corresponding to 500 TWh for the EU. A major increase in energy consumption for air-conditioning is expected to occur between 2006 and 2030 (Weiss and Biermayr 2009).

The impact of cooling in Europe is increasing, yet substantive data, statistics, and prognoses on the continent's current cold market and energy consumption remain scarce. At present, the use of energy for comfort cooling is to a high degree unknown on an aggregated EU level. In contrast to the heat market, estimations and predictions for cooling are more complex (ECOHEATCOOL 2006), since its electricity use is usually embedded in the buildings' total electricity consumption. Usage is spread across a range of electricity consumption equipment and is very rarely monitored on an aggregated level. Moreover, aggregated benchmarking information is not being systematically collected. Since the estimation of electricity use is also built up from various sources (such as chillers, auxiliary equipment, re-cooling systems, and even ventilation systems), it is a complex task to monitor and to analyze the cooling energy use in buildings. Data for the services sector's energy consumption are less detailed and complete than for the residential sector (Weiss and Biermayr 2005). Reliable estimations on cooling-energy consumption have to be made either by means of the cooling capacity of sold or installed equipment and the assumptions of the cooling demand or by means of costly surveys. However, all reliable sources document a strong increase in the cooled and air-conditioned built environment in Europe. A continuous growth is expected in both the residential and the service sectors.

So far, in most European countries, the amount of energy required for heating is much greater than the one used for space cooling. However due to high internal loads, the proliferation of fashionable glass façades, to thermal insulation and rising standards of comfort, the cooled area is steadily increasing. Events like the extraordinarily hot summer of 2003 are accelerating this trend and steadily rising mean annual temperatures are increasing the specific energy demand for space cooling (Aebischer et al. 2007). Considering air-conditioning in the residential sector, a correlation between the market saturation and the climate can be observed. The nondomestic market probably has different dynamics, but there is little reliable information on these (Riviere 2008). The current level of sales suggests that climate is less influential, with relatively high levels of sales (relative to residential use) in moderate climates. In the USA, Japan, and (as far as can be ascertained) in Europe, market penetration into nondomestic buildings is higher than into dwellings (the USA is 80 % commercial and 65 % residential, Japan is 100 % commercial and 85 % residential, Europe is 27 % commercial, and 55 % residential) (Riviere 2008).

The European market for air-conditioning is relatively young and still growing substantially. The installed stock is far from the saturation levels seen in other parts of the world and the sales figures show no sign of approaching market saturation. There are very few relevant statistics on the stock of installed products, but rather on sales—though not comprehensive (Riviere 2008).

The relative growth of the electricity demand is largest in southern countries due to general comfort requirements calling for more cooling, a trend which is slightly reinforced by the changing climate. Although the share in electricity for cooling purposes will increase in all countries, it does so at a significantly higher level in southern Europe, by about 45 % (Jochem and Schade 2009). Because of a strongly differing growth rate across EU member states, the relative share of the total cooled floor-area of EU countries such as France or Germany, which was large in the 1980s, has diminished in the 1990s. The high growth rate in central air-conditioning systems installed in Italy and Spain means that these countries now account for more than 50 % of the EU market.

In the building sector, decentralized air-conditioning units dominate the distribution systems and district cooling accounts with 1–2 % of the cooling market. In 2007, 500 million Euros were generated by selling cooling, air-conditioning, and ventilation systems in Germany. Therefore, Germany is the European market leader, followed by Italy with a revenue of 460 million Euros. Air-conditioning systems contribute to the major part of the revenue (50,000 units) with approx. 500 million Euros, followed by water-based chillers (7,000 units) with 150 million Euros (Chillventa 2009).

In the EU, the energy efficiency of air-conditioning systems is not a criterion that presently plays any major role in their design or installation process; the efficiency improvements that do occur rather tend to happen haphazardly. Air-conditioning constitutes a rapidly growing electrical end use in the European

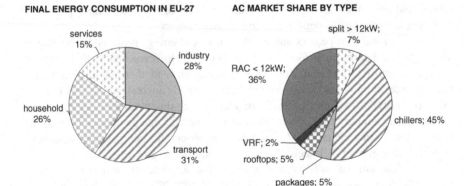

Fig. 1.1 *Left* Share of total final energy consumption [%] distributed to the major energy service sectors in EU-27 for 2006 (Weiss and Biermayr 2005). *Right* AC market share by AC type expressed in terms of newly installed cooled area in EU buildings in 1998 (Adnot 2003)

Union, yet the possibilities for improving its energy efficiency have not been fully investigated. The average Energy Efficiency Ratio (EER) is about 3.57 for water-based systems, whereas it is 2.52 for systems with air as a rejection medium under conditions of a testing standard (Adnot 2003). For the electricity consumption to meet the cooling load, it is assumed that system losses (such as auxiliary-heat supplementary load, distribution losses, suboptimal control, etc.) account for 25 % of the load. The aggregated seasonal energy efficiency ratio (SEER) for cooling is expected to be 2. Additional electricity consumption for air-handling units, pumps, and other auxiliaries not taken into account in the SEER value is at 25 % of the electric consumption (Riviere et al. 2010) (Figs. 1.1, 1.2).

1.2 Technologies and Concepts for Cooling Nonresidential Buildings

At present, there is no unambiguous classification for cooling strategies or terminology. Cooling strategies for nonresidential buildings may be distinguished as follows:

- *Passive Low-Exergy cooling*: Passive cooling strategies refer to technologies or building design features that cool the building space or prevent the building from overheating without any energy consumption, i.e., energy-consuming components such as fans or pumps are not used. Passive cooling techniques encompass heat and solar protection, heat modulation and dissipation: solar shading, high-quality building envelope, passive use of solar-heating gains, daylighting concepts, sun-protection glazing, static solar shading devices, heavyweight building construction, moderate ratio of glass to façade, and natural ventilation through open windows (Santamouris 2007; Pfafferott 2004).

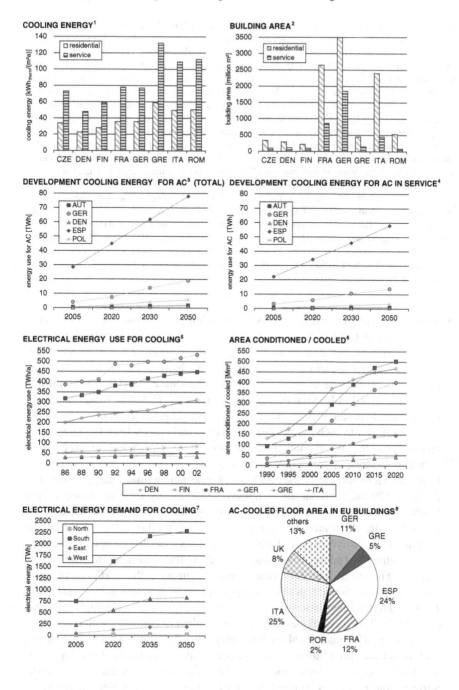

◀ **Fig. 1.2** Current and future cooling market in Europe. **Chart 1** Calculated specific cooling-energy demand [kWh$_{therm}$/m^{2}a] for selected European countries, considering the residential and service-building sector (ECOHEATCOOL 2006). The calculation method includes: frequency of outdoor temperatures, national electricity-demand variations, and specific market information from international databases, international statistical reports and commercial-market reports. **Chart 2** Total building area of the residential and service sector for selected European countries [million m^2] (ECOHEATCOOL 2006). **Chart 3** Future development of energy consumption for air-conditioning [TWh$_{el}$] (Weiss and Biermayr 2005); results of the expected development of energy consumption (electricity) for air-conditioning. "The calculation of the long-term energy consumption for air-conditioning is based on a bottom-up model and considers both impacts, the current diffusion of technology and the influence of higher cooling demand due to increasing cooling-degree-days from global warming" (Weiss and Biermayr 2005). **Chart 4** Energy consumption for air-conditioning in the service sector [TWh$_{el}$] (Weiss and Biermayr 2005). **Chart 5** Annual electrical energy use for cooling [TWh$_{el}$] available on the domestic market for the period from 1986 to 2002 (ECOHEATCOOL 2006). **Chart 6** Area air-conditioned in each country and year. The cooled area is estimated by combining manufacturer statistics (capacities, number of pieces) and national statistics [square meters cooled] (Adnot 2003). **Chart 7** Electrical-energy demand for cooling in four European regions (EU-27 + 2) [TWh$_{el}$] (Jochem and Schade 2009). **Chart 8** National shares of installed central air-conditioning floor area in EU buildings in 1998 (Adnot 2003)

- *Active Low-Exergy Cooling*: In IEA ECBCS Annex 37, 'low-exergy (or LowEx) systems' are defined as heating or cooling systems that allow the use of low-valued energy as their energy source, e.g., environmental heat sources and sinks, waste heat, etc. In practice, this means systems that provide heating or cooling energy at a level close to room temperature (Ala-Juusela 2003). Environmental energy is defined as low-temperature heat source (4–15 °C) in winter and high-temperature heat sink (15–25 °C) in summer, being provided in close proximity to the building site such as surface-near geothermal energy from the ground and groundwater, the use of rainwater and ambient air. Borehole heat exchangers, ground collectors, energy piles, earth-to-air heat exchangers and ground-water wells are technologies to harvest surface-near geothermal energy down to a depth of 120 m. Ambient air is utilized naturally, by opening windows and ventilation slats, or mechanically, by supply and/or exhaust air systems. In cooling mode, environmental heat sinks are often used directly or with a heat exchanger. Electrical energy (high exergy) is only needed to operate the auxiliary equipment (pumps and fans) as well as the measurement and control systems. Though a clear definition is missing, many authors, HVAC planners and companies define integrated concepts with cooling temperatures higher than 16 °C and hybrid ventilation systems as "low-energy cooling concepts" (Babiak et al. 2007). These techniques may only be applied to buildings with a low heat-flow density, which is typical for low-energy buildings only. Hence, low-energy cooling is based on a building concept which allows for reduced cooling loads.
- *Mechanical Cooling*: Mechanical cooling includes systems that use common refrigeration processes applied for air-conditioning (air-based systems) or radiant (i.e., water-based) cooling. Most of the cold production for air-conditioning for buildings is generated with vapor compression machines. In the

evaporator, the refrigerant evaporates at a low temperature. The heat extracted from the external water supply is used to evaporate the refrigerant from the liquid to the gas phase. The external water is cooled down or—in other words—cooling power becomes available. The key component is the compressor, which compresses the refrigerant from low pressure at a low temperature to higher pressure (at a high temperature) in the condenser. Electrical energy (high exergy) is consumed by the motor used to drive the compressor (Henning 2004).

- *Thermally Driven Cooling*: Thermally driven chiller-based and desiccant systems are key solutions for solar-assisted air-conditioning. The process principle is the same as described before, but the driving energy is heat in the sense of a thermally driven process. Most common types of thermally driven chillers are absorption and adsorption chillers. The working principle of an absorption system is similar to that of a mechanical compression one with respect to the key components of evaporator and condenser. A vaporizing liquid extracts heat at a low temperature (cold production). The vapor is compressed to a higher pressure and condenses at a higher temperature (heat rejection). The compression of the vapor is accomplished by means of a thermally driven "compressor" consisting of the two main components of absorber and generator. The heat required can be supplied, for instance, by direct combustion of fossil fuels, by waste heat or solar energy. Instead of absorbing the refrigerant in an absorbing solution, it is also possible to adsorb the refrigerant on the internal surfaces of a highly porous solid. This process is called "adsorption." Typical examples of working pairs are water/silica gel, water/zeolite, ammonia/activated carbon or methanol/activated carbon. In absorption machines, the ability to circulate the absorbing fluid between absorber and desorber results in a continuous loop. On adsorption machines, the solid sorbent has to be alternately cooled and heated in order to be able to adsorb and to desorb the refrigerant. Operation is therefore periodic in time (Henning 2004). More details on thermally driven chillers are given in "Solar-Assisted Air-Conditioning in Buildings—A Handbook for Planners" (Henning 2004).

- *District Cooling*: District cooling is a system in which chilled water for space and process cooling is distributed in pipes from a central cooling plant to buildings. A district cooling system contains three major elements: the cooling source, a distribution system and customer installation. Chilled water is generated by compressor-driven chillers, absorption chillers or other sources like ambient cooling or "free cooling" from lakes, rivers, or oceans. District heat may be the heat source for absorption chillers, but with today's technique, only if there is waste heat available. If the heat from the power generation process wasn't being used in summer, it could be economically converted and used to produce cooling. The production from a centralized facility allows for improvements in energy conservation. The generation of cooling may be a mix of several energy sources, for example, chillers and free cooling. Cooling generation may also be configured with thermal storage in order to reduce chillers' equipment requirements and lower operating costs by shifting the peak load to off-peak times. The successful implementation of district cooling systems depends largely on the ability of the system to obtain high-temperature differences between supply and return water. The significant installation costs associated with a central-distribution piping system and the physical operating limitations (e.g., pressures and temperatures) of district energy systems require careful scrutiny with the design options available for new and existing buildings' HVAC systems. This is crucial to ensure that the central district cooling systems can operate with a reasonable size of distribution piping and pumps in order to minimize the pumping-energy requirements (LowEx IEA ECBCS Annex 37).

The monitored and analyzed nonresidential buildings in the framework of the guidebook employ passive and active low-exergy cooling concepts as well as mechanical cooling strategies. Table 1.1 presents a suggested categorization of cooling in nonresidential buildings and describes the heat sinks usually applied to the systems. The guidebook of IEA ECBCS Annex 37 presents a very comprehensive description of LowEx-technologies—for harvesting environmental energy and for the delivery of heating/cooling energy to the building space—for heating and cooling applied in nonresidential and residential buildings.

1.3 Building Requirements

In the past few decades, the envelope of modern buildings has improved significantly. Therefore, buildings with a comparatively low heating and cooling demand—the cornerstone of a sustainable energy concept—can be realized in mid-European climates. Those buildings aim at establishing a pleasant interior environment without costly building service equipment and excessive energy use. While the heating demand in nonresidential buildings could be reduced significantly, the cooling demand is growing because of increased internal loads by office appliances and increased glazed areas on modern commercial buildings. This trend has been amplified by warmer summers in many areas and an increased demand for comfort.

Cooling demand for comfort purposes in buildings is mainly due to climatic conditions. Other important factors are building standards, the cooling system installed, and occupant behavior (ECOHEATCOOL 2006):

- Regional climatic conditions: temperature and humidity differences depending on geographical position. The predominating factor is usually the outdoor air temperature.
- Urban climatic conditions: the climate in densely built areas can differ from the surrounding climate, for example in temperature, wind speed, and humidity.
- Building design: the architectural and structural design features of the building have a strong impact on its indoor climate (building layout, insulation, window

Table 1.1 Categorization of cooling systems into (I) passive LowEx-cooling, (II) active LowEx-cooling, (III) mechanical cooling and (IV) thermally driven cooling see also Table 5.1 for technical application

	(I) Passive LowEx cooling			(II) Active LowEx cooling				
	Structural design	Air-based systems		Air-based systems			Water-based systems	
	De-/Central	Decentral	Central	Decentral	Central		Central	
					w/ cooling	w/o cooling	Direct cooling	Indirect cooling
Cooling generation	Quality of building shell (insulation) Air-tightness Static/Movable solar-shading devices Solar glazing Window/Façade ratio	Windows Ventilation slats	Solar chimney Atrium	Façade-ventilation unit	Exhaust-air system Exhaust-/Supply-air system	Exhaust-/Supply-air system with earth-to-air heat exchanger Direct evaporative cooling Indirect evaporative cooling	Borehole heat exchanger w/o heat exchanger	Borehole heat exchanger Groundwater well Cooling tower w/ heat exchanger
Heat rejection	None	Ambient air	Ambient air	Ambient air	Ambient air	Heat recovery Ambient air Ground	Ground	Ground Groundwater Ambient air

(continued)

Table 1.1 (continued)

	(I) Passive LowEx cooling			(II) Active LowEx cooling				
	Structural design	Air-based systems		Air-based systems			Water-based systems	
	De-/Central	Decentral	Central	Decentral	Central		Central	Indirect cooling
					w/ cooling	w/o cooling	Direct cooling	
Temperature-level heat sink	–	Variable	Variable	Variable	Variable	8–18 °C	8–18 °C	8–18 °C
Air treatment	None	None	None	None	None	None	None	None
Distribution system	None	None	None	Fan	Air-duct system, fan	Air-duct system, fan	Hydronic system	Hydronic system
Delivery system	None	None	None	None	None	None	TABS Ventilation system	TABS Ventilation system
Temperature-level delivery system	–	Variable (ambient air)	Variable (ambient air)	Variable (ambient air)	Variable (ambient air)	16–24 °C	16–22 °C	16–22 °C
Final energy	None	None	None	Low l aux. energy	Low l aux. energy	Low l aux. energy	Low l aux. energy	Low l aux. energy
Efficiency (SPF)	–	–	–	Low	SPF 4	SPF 4	SPF 15–20	SPF 15–20
Costs	Low–Moderate	Low	Low	Moderate	Moderate	Moderate	Intensive	Intensive

(continued)

Table 1.1 (continued)

	(III) Mechanical cooling					(IV) Thermally driven cooling			
	Air-based systems		Water-based systems			Heat transformation		Closed cycle	
	Decentral	Central	Central		Water-to-water	Desiccant (open cycle)			
	Air-to-air	Air-to-air	Water-to-air	Air-to-water	Water-to-water	Liquid sorbent	Solid sorbent	Absorption	Adsorption
Cooling generation	Split unit	Roof-top split unit; Rev. heat pump; VRF	Rev. heat pump	Rev. heat pump; Chiller	Rev. heat pump; Chiller	Counterflow; Absorber	Dehumidifier; Rotor; Fixed-bed process	Liquid sorbent; Water/lithium bromide	Solid sorbent adsorption (water/silica gel)
Heat rejection	Ambient air	Ambient air	Ground; Groundwater; Ambient air	Ambient air	Ground; Groundwater	Solar energy; District heat	Solar energy; District heat	Water/lithium bromide	Solar energy; District heat
Temperature-level heat sink	Variable	Variable	8–18 °C	Variable	8–18 °C	None	None	35–60 °C	30–40 °C
Air treatment	Dehumid.	Dehumid.	Dehumid.	Dehumid.	Dehumid.	Dehumid.	Dehumid.	None	None
Distribution system	None	Air-duct system	Air-duct system	Hydronic pipe system	Hydronic pipe system	Air-duct system	Air-duct system	Hydronic pipe system	Hydronic pipe system

(continued)

Table 1.1 (continued)

	(III) Mechanical cooling				(IV) Thermally driven cooling			
	Air-based systems		Water-based systems		Heat transformation			
					Desiccant (open cycle)		Closed cycle	
	Decentral	Central	Central	Fan-coil unit	Desiccant (open cycle)	AHU	Fan-coil unit	Induction unit
Delivery system	None	Fan-coil unit Induction unit AHU	Fan-coil unit Induction unit AHU TABS	Induction unit AHU TABS	AHU	AHU	Fan-coil unit Induction unit AHU TABS	Induction unit AHU TABS
Temperature-level delivery system	14–20 °C	14–20 °C	14–22 °C	14–22 °C	14–22 °C	14–22 °C	6–12 °C	13–18 °C
Final energy	High	High	High	High	Low, waste heat	Low, waste heat	Moderate	Low, waste heat
Efficiency (SPF)	SPF 1.9–3.3	SPF 3	SPF 2.5–4.0	SPF 3–6	SPF > 10 evaporative cooling	SPF > 10 evaporative cooling	SPF 0.6–1.3, thermal energy only	SPF 0.6–0.8
Costs	Low	Moderate	Moderate	High	Low	Low	Moderate	Moderate

orientation, shading, ventilation, daylighting concept, and micro-climate around the building).

- Building use: internal heat gains from occupants, lighting, and equipment account for cooling demand.
- Comfort requirements and use: working hours, vacation period, and the required indoor temperature have a major impact on cooling demand and consumption.

References

Adnot J (ed) (2003) Energy efficiency and certification of central air conditioners (EECCAC). Final report, vol 1–3. Study for the D.G. Transportation-Energy (DGTREN) of the Commission of the E.U, Paris

Aebischer B, Catenazzi G, Jakob M (2007) Impact of climate change on thermal comfort, heating and cooling energy demand in Europe. In: Proceedings ECEEE summer studies. Saving energy—just do it!, European Council for an Energy Efficient Economy, La Colle Sur Loup, 4–9 June 2007

Ala-Juusela M (ed) (2003) LowEx guidebook: low-exergy systems for heating and cooling of buildings. Guidebook to IEA ECBCS annex 37. International energy agency—energy conversation in buildings and community systems programme. ECBCS Bookshop, Birmingham

Babiak J, Olesen BW, Petrás D (2007) Low temperature heating and high temperature cooling. REHVA Guidebook 7. Rehva, Brüssels

Brown MA, Southworth F, Stovall TK (2005) Towards a climate-friendly built environment. Prepared for the Pew Center on global climate change. Oak Ridge National Laboratory. http://www.pewclimate.org/technology-solutions/pubs. Accessed Dec 2013

Chillventa (2009) http://www.chillventa.de/en/

COM (2008) Communication from the commission to the European Parliament, the Council, The European Economic and Social Committee and the Committee of the Regions. Europe's climate change opportunity., Brussels, 23 Jan 2008

ECOHEATCOOL (2006) The European cold market, final report, work package 2, Ecoheatcool and Euroheat & Power 2005–2006. Euroheat & Power, Brussels

EPBD (2010) Directive 2010/31/EU of the European Parliament and of the Council of 19 May 2010 on the energy performance of buildings. OJ L 153:13–35. http://eur-lex.europa.eu/

EUR-Lex (2002) Directive 2002/91/EC of the European Parliament and of the Council of 16 December 2002 on the energy performance of buildings. Official Journal of the European Communities. L 1/65. http://eur-lex.europa.eu

European Environment Agency (2006) Greenhouse gas emission trends and projections in Europe. Copenhagen. http://www.eea.europa.eu/publications/. Accessed Dec 2013

Henning H-M (ed) (2004) Solar-assisted air-conditioning in buildings—a handbook for planners, 1st edn. Springer, Wien

Jochem E, Schade W (2009) ADAM 2-degree scenario for Europe—policies and impacts. Deliverable D3 of work package M1. ADAM Adaptation and mitigation strategies: supporting European climate policy. Fraunhofer Institute for Systems and Innovation Research (Fraunhofer-ISI), Karlsruhe

McKinsey & Company (2007) Reducing U.S. greenhouse gas emissions: how much at what cost. http://www.mckinsey.com/client_service. Accessed Dec 2013

Pfafferott J (2004) Enhancing the design and operation of passive cooling concepts. Dissertation, University of Karlsruhe. Fraunhofer IRB Verlag, Stuttgart

Riviere P (ed) (2008) Preparatory study on the environmental performance of residential room conditioning appliances. Final report of ECODESIGN project, contract TREN/D1/40-2005/ LOT10/S07.56606

Riviere P, Adnot J, Spadaro J, Hitchin R, Pout C, Kemna R, van Elburg M, van Holsteijn R (2010) Sustainable industrial policy—building on the ecodesign directive—energy-using product group analysis. Final report of ECODESIGN project, contract TREN/B1/35-2009/ LOT6/S12.549494

Santamouris S (ed) (2007) Advances in passive cooling. EarthScan, London

Stern N, Peters S, Bakhshi V, Bowen A, Cameon C, Catovsky S, Crane D, Cruickshank S, Dietz S, Edmonson N, Garbett S-L, Hamid L, Hoffman G, Ingram D, Jones B, Patmore N, Radcliffe H, Sathiyarajah R, Stock M, Taylor C, Vernon T, Wanjie H, Zenghelis, D (2006) Stern review: the economics of climate change. HM Treasury London. http://www.hm-treasury.gov. uk/stern_review_report.htm. Accessed July 2011

Weiss W, Biermayr P (2005) The solar thermal potential in Europe. Final report of RESTMAC project. AEE—Institute for Sustainable technologies, Vienna

Weiss W, Biermayr P (2009) Potential of solar thermal in Europe. Final report of RESTMAC. Bruxelles, Belgium

Chapter 2
Thermal Indoor Environment

Abstract Room temperature and indoor air quality have a strong impact on the overall satisfaction with the thermal environment. Responses to our thermal indoor environment have a considerable effect on health, comfort, and performance. Formal methods have been developed to design the interior environment. Thermal comfort takes both global and local parameters as well as static and dynamic aspects into consideration.

The thermal indoor environment is a composition of many and diverse aspects. Hence, the perspective on thermal comfort may change its evaluation by occupants (Corgnati et al. 2011).

2.1 Human Responses to the Thermal Environment

Responses to our thermal indoor environment have a considerable effect on health, comfort, and performance. There has been considerable scientific investigation into these responses and formal methods have been developed to design and to develop the interior environment. Existing methods for the evaluation of the general thermal state of the body, both in comfort and under heat- or cold-stress considerations, are based on an analysis of the heat balance for the human body. Under cool to thermo-neutral conditions, heat gain is balanced by heat loss, no heat is stored, and body temperature equilibrates, that is:

$$S = M\text{-}W\text{-}C\text{-}R\text{-}E_{sk}\text{-}C_{res}\text{-}E_{res}\text{-}K \text{ in W/m}^2 \tag{2.1}$$

where:

S Heat storage in the human body;
M Metabolic heat production;
W External work;
C Heat loss by convection;
R Heat loss by radiation;

D. E. Kalz and J. Pfafferott, *Thermal Comfort and Energy-Efficient Cooling of Nonresidential Buildings*, SpringerBriefs in Applied Sciences and Technology, DOI: 10.1007/978-3-319-04582-5_2, © The Author(s) 2014

E_{sk} Evaporative heat loss from skin;
C_{res} Convective heat loss from respiration;
E_{res} Evaporative heat loss from respiration;
K   Heat loss by conduction.

The four environmental factors influencing this heat balance are: air and mean radiant temperature (°C), air speed (m/s), and partial water vapor pressure (Pa). The three personal variables are: metabolic heat production due to the activity level (W/m^2 or met), the thermal resistance of clothing (clo or m^{2}K/W), and the evaporative resistance of clothing (m^{2}Pa/W). These parameters must be in balance so that the combined influence will result in a thermal storage equal to zero. A negative thermal storage indicates that the environment is too cool, and vice versa. In order to provide thermal comfort, the mean skin temperature also has to be within certain limits and the evaporative heat loss must be low.

Human responses to the thermal environment and to internal heat production serve to maintain a narrow range of internal body temperatures of 36–38 °C. The human body has a very effective thermoregulation system, which uses blood flow for heat transport (high blood flow: enhanced heat dissipation—low blood flow: reduction of heat losses) with the hypothalamus acting as its main "thermostat."

There are two categories of human responses to the thermal environment: voluntary or behavioral responses, and involuntary or physiological autonomic responses. Voluntary or behavioral responses generally consist of avoidance or reduction of thermal stress through modification of the body's immediate environment or of clothing insulation. Physiological responses consist of peripheral vasoconstriction to reduce the body's thermal conductance and increased heat production by involuntarily shivering in the cold, and of peripheral vasodilation to increase thermal conductance and secretion of sweat for evaporative cooling in hot environments. Autonomic responses are proportional to changes in internal and mean skin temperatures. Physiological responses also depend on the point in a diurnal cycle, on physical fitness, and on the sex of the individual. Behavioral responses rely on thermal sensations and discomfort. The latter appears to be closely related to the level of autonomic responses so that warm discomfort is closely related to skin wetness and cold discomfort similarly relates to cold extremities and shivering activity.

However, there is no physiological acclimatization to cold environments; the most common way to compensate for cold environments is behavioral adaptation by clothing adjustment. In warm environments, sweating is a very efficient way of losing heat. However, the sweat rate is limited, as well as how much a person can sweat during a day. Clothing, posture, and reduced activity are all behavioral ways of adapting to hot environments. Studies have also shown that people's expectations may change and influence their acceptability of the thermal environment. Besides the general thermal state of the body, a person may find the thermal environment unacceptable or intolerable if the body experiences local influences from asymmetric radiation (opposite surfaces with high temperature differences, solar radiation on single parts of the body, air velocities, vertical air-temperature differences or contact with hot or cold surfaces (floors, machinery, tools, etc.).

In existing standards, guidelines or handbooks, different methods are used to evaluate the general thermal state of the body in moderate, cold, and hot environments; but all are based on the heat balance and listed factors (EN ISO 7730:2005 2005; DIN EN ISO 11079:2007 2007).

Due to individual differences, it is impossible to specify a thermal environment that could satisfy everybody. There will always be a share of dissatisfied occupants. However, it is possible to specify environments that are likely to be acceptable for a certain percentage of the occupants. If they have some kind of personal control (change of clothing, setting of room temperature in a single office, increase of air velocity, change of activity level and posture), the overall satisfaction with the environment will increase significantly and every occupant may be satisfied. Due to local or national priorities, technical developments and climatic regions, a higher thermal quality (less dissatisfied occupants) or a lower one (more dissatisfied occupants) may be sufficient in some cases.

2.2 Health and Individual Performance

Besides influencing people's comfort, the thermal environment may also affect their health and performance:

- Extreme cold or hot environments are of high risk to the human body (heat stroke, frostbites, etc.). However, if thermal indoor conditions are less extreme, raised room temperatures have been associated with an increased prevalence of symptoms typical for the Sick Building Syndrome (SBS) or nonspecific, yet building-related symptoms such as headache, chest tightness, difficulty in breathing, fatigue, irritation of eyes and mucous membranes—all of which may be alleviated after the individual leaves the building (WHO 1983).
- Thermal conditions can affect productivity and work performance through several mechanisms, such as the following: thermal discomfort distracts attention and generates disorders that increase maintenance costs. SBS symptoms have a negative effect on mental work. Cold conditions lower finger temperatures and thus have a negative effect on manual dexterity. Rapid temperature swings have the same effects on office work as slightly raised room temperatures, while slow temperature swings cause discomfort that can lower concentration and increase disorders. Vertical thermal gradients reduce the perceived air quality or lead to a reduction in room temperature which then causes troubles of cold at floor level.
- The hypothesis that thermal conditions within the thermal comfort zone do not necessarily lead to optimum work performance is supported by the results of several studies. They showed that subjects performed best at a temperature

lower than thermal neutrality. However, a strong relationship between thermal sensation and relative office-work performance, based on a statistical analysis of data from laboratory and field measurements, is discussed with controversy. Some authors establish quantitative relations between indoor environmental quality and work performance and even derive a model that integrates the economic outcome of improved health and performance into building cost-benefit calculations, in conjunction with initial, energy and maintenance costs. Other authors interpret the results from these field studies from the users' perspective and conclude that their performance is strongly related to their satisfaction with the thermal environment. And user satisfaction is related to both their expectation and sensation of the thermal environment.

In the context of this guidebook, we assume that low-energy cooling improves the indoor environment, hence reduces negative health effects and improves the work performance. However, these effects have not been quantified or even economically evaluated.

2.3 Criteria for Thermal Comfort

Different physical parameters affect physiological reactions to the environment. Thus, these parameters (air, radiant and surface temperature, air velocity and humidity) are also the basis for defining criteria for an acceptable thermal environment. The criteria result in requirements for general thermal comfort (PMV/PPD index or operative temperature) and for local comfort disturbance (i.e., draft, radiant asymmetry, vertical air temperature differences and requirements on surface-temperature differences). They can be found in international standards and guidelines such as EN ISO 7730:2005 (2005), CEN/CR 1752 (2001), EN 15251:2007-08 (2007), and ASHRAE 55:2004-04 (2004), or in their national derivate respectively.

Operative Room Temperature. The most important criterion for the thermal environment is the operative temperature. As a sufficient approximation for most cases, it can be calculated as the arithmetic mean of the air temperature and the mean radiant temperature of surrounding surfaces in an occupied zone. Air temperature refers to the average value of the temperature in space and time in an occupied zone (ASHRAE 55:2004-04 2004). For a first general thermal comfort evaluation, a simplified calculation of the mean radiant temperature can be carried out, with surface temperatures weighted by the different surface areas.

Parameters for Local Discomfort. Besides the operative temperature, there are further temperature-related criteria to describe the thermal environment, particularly to assess local discomfort. The temperature asymmetry is also based on the radiant temperature of surfaces and is defined as the temperature difference between either two vertical (wall) or horizontal (ceiling and floor) surfaces. Another criterion is the absolute surface temperature, mainly important for the floor to which the body has constant contact for long periods in many situations. Finally, the stratification of the air temperature has to be taken into account as a criterion for local comfort. Accepted temperature ranges for these three criteria can be found in EN ISO 7730:2005 (2005) and ASHRAE 55:2004-04 (2004).

Humidity. Humidity is addressed only as a boundary condition for general comfort [an upper limit is given in (ASHRAE 55)].

Air Velocity. Air velocity can be experienced either as draft sensation or may lead to improved thermal comfort under warm conditions. It may occur due to enforced air movement (open window/door, air outlet of ventilation system) or to buoyancy effects (air falling down along a cold window surface). Allowable air velocities and acceptable limits for draft rates—in terms of predicted percentage of people dissatisfied with draft—are summarized in (ISO 7730:2005 2005) and (ASHRAE 55:2004-04 2004). ISO 7730:2005 (2005) describes an allowance for higher air velocities in order to offset an increased operative room temperature, which was adopted by ASHRAE 55:2004-04 and in EN 15251:2005-07 standards.

2.4 Overall Satisfaction with the Thermal Environment

Further environmental parameters, e.g., air quality, visual or aural environment, can interact with the thermal environment and therefore influence thermal comfort or overall satisfaction in a space.

For most parameters describing the thermal environment, relationships between the parameter itself and a predicted percentage of people rating the indoor condition as (un)acceptable are established. People may be dissatisfied due to general thermal comfort and/or local thermal comfort parameters. However, there is no method for combining these percentages of dissatisfied persons to give a good prediction of the total number of occupants deeming the thermal environment unacceptable.

In comparison to the thermal environment, there is a large number of criteria and requirements for other indoor environment qualities such as air quality, visual and aural comfort. On the one hand, there is a possible physical interference of the different comfort requirements, e.g., for daylight and resulting solar heat gains through windows or recommended ventilation rates and noise from outside through open windows. On the other hand, the various comfort criteria have an impact on the (overall) occupants' satisfaction with the workplace and probably on thermal comfort. They also include social and architectural aspects related to a specific workspace. As there is not enough proof for quantitative correlations, their evaluation is only possible through a direct assessment after the building went under operation.

Figure 2.1 exemplarily shows a survey result for German office buildings. The subjective votes on satisfaction levels with different environmental parameters were given with respect to their relevance for the occupants' overall satisfaction with their workplaces (Gossauer and Wagner 2008).

Fig. 2.1 Matrix with relevant satisfaction parameters from a field study in 17 German office buildings. The parameters are weighted by their correlation coefficients against the overall satisfaction with the workplace (importance of the parameter). For details of the survey, see (Gossauer and Wagner 2008)

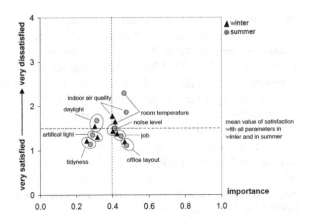

The lower left field shows parameters with high satisfaction levels but the weighting calculation shows that they are less important for the general satisfaction with the workplace. In the lower right field, occupants are satisfied with the parameters that are more important for general satisfaction levels. The upper left square shows parameters with higher dissatisfaction but less importance for general satisfaction whereas parameters in the upper right combine higher dissatisfaction levels with higher importance for the general satisfaction with the workplace.

References

ASHRAE Standard 55-2004 (2004) Thermal environmental conditions for human occupancy. American Society of Heating, Refrigerating and Air-Conditioning Engineers Inc., Atlanta

CEN/CR 1752 (2001) Ventilation for buildings—design criteria for the indoor environment. European Committee for Standardization, Brussels

Corgnati SP, Gameiro da Silva M, Ansaldi R, Asadi E, Costa JJ, Filippi M, Kaczmarczyk J, Melikov AK, Olesen BW, Popiolek Z, Wargocki P (2011) Indoor climate quality assessment—evaluation of indoor thermal and indoor air quality. Rehva guidebook 14. Rehva, Brussels

DIN EN ISO 11079:2007-12 (2007) Ergonomics of the thermal environment—determination and interpretation of cold stress when using required clothing insulation (IREQ) and local cooling effects. Beuth, Berlin

EN 15251:2007-08 (2007) Criteria for the indoor environment. Beuth, Berlin

EN ISO 7730:2005 (2005) Ergonomics of the thermal environment—analytical determination and interpretation of thermal comfort using calculation of the PMV and PPD indices and local thermal comfort criteria. Beuth, Berlin

Gossauer E, Wagner A (2008) Occupant satisfaction at workplaces—a field study in office buildings. In: Proceedings of windsor conference on air-conditioning and the low carbon cooling challenge, London Metropolitan University, Windsor, July 2008

World Health Organization (WHO) (1983) Indoor air pollutants: exposure and health effects. EURO Reports and Studies No. 78, WHO Regional Office for Europe, Copenhagen

Chapter 3
Standards on Thermal Comfort

Abstract International standards give criteria for thermal comfort based on the evaluation of room temperatures and their deviation from the comfort temperature. The static approach to thermal comfort is derived from the physics of heat transfer and combined with an empirical fit to sensation. It defines constant comfort temperatures for the summer and winter period considering different clothing. The adaptive comfort model considers the thermal sensation of the occupants and different actions in order to adapt to the thermal environment as well as variable expectations with respect to outdoor and indoor climate. The comfort temperature is dependent on the outdoor temperature. Though standards clearly define the static approach as general criterion and the adaptive approach as optional approach for naturally ventilated buildings only, the application in buildings with lowenergy cooling and strong users' impact on the indoor environment is critically reviewed.

Human thermal comfort is defined as the state of mind that expresses satisfaction with the surrounding environment (ASHRAE 55:2004-04 2004). Thermal comfort is achieved when thermal equilibrium is maintained between the human body and its surroundings and the person's expectations of the surrounding conditions are satisfied. Occupant satisfaction was first considered in the 1980s, when it was found that some chronic illness was building-related (e.g., reported symptoms like lethargy, headaches, dry eyes, and dry throat) (Bordass et al. 2001a). Current national and international regulations draw on diverse—and partly controversial—results from thermal-comfort studies carried out in laboratories or in the field. There are two main models to determine human thermal comfort and to predict the occupant's satisfaction with the interior conditions: (i) the *heat-balance approach* used in the standard EN ISO 7730:2005 (2005) and (ii) the *adaptive approach* described in the standards EN 15251:2007-08 (2007), ASHRAE 55:2004 (2004), and the Dutch guideline ISSO 74:2005 (Boersta et al. 2005). The discussion about thermal comfort and user satisfaction has mainly been concerned with nonresidential buildings rather than dwellings, but has implications for the residential sector.

D. E. Kalz and J. Pfafferott, *Thermal Comfort and Energy-Efficient Cooling of Nonresidential Buildings*, SpringerBriefs in Applied Sciences and Technology, DOI: 10.1007/978-3-319-04582-5_3, © The Author(s) 2014

3.1 Static Approach to Thermal Comfort

The static approach to thermal comfort (EN ISO 7730:2005) is derived from the physics of heat transfer and combined with an empirical fit to sensation (predicted mean vote and predicted percentage of dissatisfaction) (Fanger 1970). The required four environmental input variables are air and mean radiant temperature, air speed, and humidity. The two personal variables are clothing and metabolic heat production. The predicted mean vote PMV is the thermal comfort index probably most widely used for assessing moderate thermal indoor environments. It rests on the steady state heat transfer theory, obtained during a series of studies in climatic chambers, where the climate was held constant. It predicts the expected comfort vote of occupants on the ASHRAE scale of subjective warmth (-3 cold to $+3$ hot) as well as the predicted percentage of dissatisfaction (PPD) for a certain indoor condition.

ISO 7730:2005-10 | Room Temperature Versus Seasonal Ambient Air Temperature. Thermal-comfort requirements in ISO 7730 rest upon the heat-balance approach (Fanger 1970) and are distinguished into a summer and a winter season. The ranges of temperature which occupants of buildings will find comfortable are merely influenced by the characteristic heat insulation of clothing. Therefore, the defined comfort criteria are generally applicable for all rooms, independent of the building technology for heating, cooling, and ventilation (Fig. 3.1).

$$\theta_{o,c} = 24.5\,^{\circ}\text{C for summer season} \tag{3.1}$$

$$\theta_{o,c} = 22.0\,^{\circ}\text{C for winter season} \tag{3.2}$$

The criterion for thermal comfort is stipulated as an average operative room temperature of 24.5 °C for the summer and 22 °C for the winter period, with a tolerance range depending on the predicted percentage of dissatisfied occupants: ± 1.0, ± 1.5, and ± 2.5 °C (classes I, II, and III).

Fanger's thermal-comfort model requires the input variables metabolic rate and the insulation level of clothing. For the winter period, a *clo* of 1.0 is assumed, which represents typical winter clothing with long-sleeved overgarment and long trousers. The summer period is described with a *clo* factor of 0.5, representing light, short-sleeved overgarment and light pants. The prevailing ambient conditions are not considered in the model. Therefore, it is not explicit how Fanger's model refers to summer or winter conditions. Judging from field studies (Haldi and Robinson 2008), the clothing level can be reliably modeled by outdoor conditions, for example by using regressions on running mean ambient air temperature. At 15 °C, this would result in a *clo* factor of 0.7 and is suggested by the authors to be used as a distinction between the winter and summer periods.

DIN EN 15251:2012-12 (National Annex) | Room Temperature versus Daily Maximum Ambient Air Temperature. The German Annex to EN 15251 defines two comfort ranges for the summer period with reference to the maximum daily

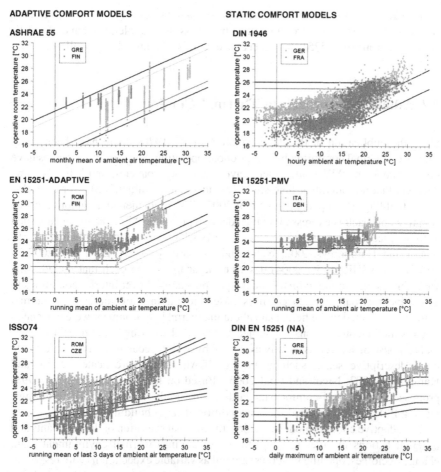

Fig. 3.1 Models for the evaluation of thermal comfort: ASHRAE 55, DIN 1946, EN 15251-Adaptive, EN 15251-PMV, ISSO 74 and DIN EN 15251 (NA) (from *upper left*) for eight low-energy buildings in Europe (for description of the buildings, see Sect. 6.1). Hourly measured operative room temperature portrayed over ambient air temperature. Reference value for ambient air temperature differs according to the comfort model

ambient air temperature—i.e., below or above a maximum temperature value of 32 °C (Fig. 3.1).

$$\theta_{o,c} = 24.5\,^{\circ}\mathrm{C} \quad \text{for} \quad \theta_{e,d,\max} \leq 16\,^{\circ}\mathrm{C} \tag{3.3}$$

$$\theta_{o,c} = 22\,^{\circ}\mathrm{C} + 0.25 \cdot (\theta_{e,d,\max} - 16\,^{\circ}\mathrm{C}) \quad \text{for} \quad 16\,^{\circ}\mathrm{C} < \theta_{e,d,\max} \leq 32\,^{\circ}\mathrm{C} \tag{3.4}$$

$$\theta_{o,c} = 26\,^{\circ}\mathrm{C} \quad \text{for} \quad \theta_{e,d,\max} > 32\,^{\circ}\mathrm{C} \tag{3.5}$$

Though the German national annex allows for categories I to IV, it defines only category II for new and retrofitted office buildings, with a tolerance band of ±2 K. Diverging from EN 15251, no exceedance of this tolerance band is allowed.

3.2 Adaptive Approach to Thermal Comfort

Since the publication of the PMV equation in the 1970s, there have been many studies on thermal comfort in buildings under operation. Some of these studies have given support to PMV while others have found discrepancies, and it has become apparent that no individual field study can adequately validate PMV for everyday use in buildings (Humphreys and Nicol 2002). The fundamental assumption of the adaptive approach is expressed by the adaptive principle: "if a change occurs such as to produce discomfort, people react in ways which tend to restore their comfort" (Nicol and Humphreys 2002; Nicol and McCartney 2002). EN 15251:2007-08 (2007) and ASHRAE 55 (2004) describe the adaptive approach that includes the variations in the outdoor climate and the person's control over interior conditions to determine thermal preferences. It is based on findings of surveys on thermal comfort conducted in the field. Data about the thermal environment were correlated to the simultaneous response of subjects under real working conditions. The thermal response of subjects is usually measured by asking occupants for a comfort vote on a descriptive scale, such as the ASHRAE or Bedford scales (Nicol and Humphreys 2002; Nicol and Roaf 2005). Based on field studies, de Dear (1998) proposed new thermal-comfort standards for naturally ventilated (NV) buildings, leaving PMV as the standard for air-conditioned (AC) buildings.

The adaptive comfort model considers the thermal sensation of the occupants and different actions in order to adapt to the (changing) thermal environment (e.g., change of clothes, opening windows) as well as variable expectations with respect to outdoor and indoor climate, striving for a "customary" temperature. The underlying assumption is that people are able to act as "meters" of their environment and that perceived discomfort is a trigger for behavioral responses to the thermal environment. Although these phenomena cannot yet be described theoretically in full detail, a model was derived from results of field studies, representing limits to the operative temperature as a function of the outdoor temperature. This simplified approach also avoids difficulties occurring with the assumption of appropriate *clo* and *met* values, as has to be done with the PMV approach. They are included in the resulting accepted temperature as part of the adaptation.

DIN 1946-2:1994-01 | Room Temperature versus Hourly Ambient Air Temperature. Although a guideline for mechanical ventilation systems in nonresidential buildings, designers and planners often fall back to the former German guideline DIN 1946-2 (1994) ("Ventilation and Air-Conditioning") in order to evaluate thermal comfort. The standard defines a correlation between the room and the prevailing ambient air temperature and is motivated by the energy-saving potential of setpoints, which depends on the ambient temperature and affects a compromise between comfort and HVAC operation. This correlation results in an adaptive criterion, since DIN 1946-2 (1994) defines the range of thermal comfort depending on the prevailing, actually ambient air temperature (Fig. 3.1):

$$\theta_{o,c} = 25\,^\circ\text{C} \quad \text{for} \quad \theta_e \leq 26\,^\circ\text{C} \tag{3.6}$$

$$\theta_{o,c} = 25\,^\circ\text{C} + 1/3 \cdot (\theta_e - 26\,^\circ\text{C}) \quad \text{for} \quad \theta_e > 26\,^\circ\text{C} \tag{3.7}$$

Since May 2005, DIN 1946-2 (1994) has been substituted by the guideline DIN EN 13779:2004-09 (2004) ("Ventilation of Non-Residential Buildings—Performance Requirements for Ventilation and Air-Conditioning Systems"), which does not consider a comparable comfort criterion any longer.

EN ISO 15251:2007-8 | Room Temperature versus Running Mean Ambient Air Temperature. EN 15251 evaluates the operative room temperature in relation to the running mean of the ambient air temperature (Fig. 3.1). Again, the temperature range defining thermal comfort in summer correlates with user satisfaction: ±2.0, ±3.0 and ±4.0 °C (classes I, II, and III). The different ranges refer to the categories defined in the standard (category I: less than 6 % dissatisfied, category II: less than 10 % dissatisfied, category III: less than 15 % dissatisfied, category IV: more than 15 % dissatisfied—based on occupants' expectations on indoor climate).

$$\theta_{o,c} = 18.8\,^\circ\text{C} + 0.33 \cdot \theta_{rm} \text{ for summer season} \tag{3.8}$$

Figure 3.1 shows these operative temperature limits for nonmechanically cooled buildings. The outdoor temperature has to be calculated as a weighted running mean value, referring to the idea that most recent experiences (last one to 7 days) might be more important for the "thermal memory." The running mean ambient air temperature θ_{rm} is given as a function of the one at the previous days θ_{rm-1} and the daily mean ambient air temperature of the previous days $\theta_{e,d-1}$ with $\alpha = 0.8$.

$$\theta_{rm} = (1 - \alpha) \cdot \theta_{e,d-1} + \alpha \cdot \theta_{rm-1} \tag{3.9}$$

ISSO74:2005 | Room Temperature versus Running Mean Ambient Air Temperature. This guideline from the Netherlands was the first European standard using an adaptive approach to thermal comfort and therefore considers the thermal adaptation. The comfortable room temperature responds to the running mean ambient air temperature of the last 3 days using the same formula as EN 15251 but with a different reference temperature (Fig. 3.1).

$$\theta_{o,c} = 17.8\,^\circ\text{C} + 0.31 \cdot \theta_{rm,2.4} \text{ for summer season} \tag{3.10}$$

$$\text{with } \theta_{rm,2.4} = (\theta_{e,d} + 0.8 \cdot \theta_{e,d-1} + 0.4 \cdot \theta_{e,d-2} + 0.2 \cdot \theta_{e,d-3})/2.4 \tag{3.11}$$

The ISSO 74 algorithm is based on a meta-analysis of existing models and considers—distinguished from other standards—the users' expectations and their efficient possibility for adaptation to the indoor environment—and not the cooling technology—as an indicator for the relevant comfort model. Since 2008, this approach has also been used for the design of new buildings for the German government (BBR-Lex 2008)—divergent from the German standard DIN EN 15251 (NA).

ASHRAE 55:2002 | Room Temperature versus Monthly Mean Ambient Air Temperature. ASHRAE 55 defines thermal comfort for naturally ventilated buildings with reference to the monthly mean ambient air temperature for the adaptive model, since this is generally available from meteorological stations (Fig. 3.1).

$$\theta_{o,c} = 17.8\,°C + 0.31 \cdot \theta_{e,\text{month}} \text{ for summer season} \tag{3.12}$$

The tolerance range is determined depending on occupant satisfaction, namely ± 2.5 °C for 90 % acceptance and ± 3.5 °C for 80 % acceptance.

Most relevant PMV and adaptive comfort models with the respective reference value for ambient air temperature are illustrated in Fig. 3.1: ASHRAE-55, EN 15251 (adaptive, PMV, and German annex), ISSO 74, DIN 1946, and ISO 7730. Thermal-comfort results are presented for eight European buildings: an hourly measurement of the operative room temperature is plotted against the specific ambient air temperature. Obviously, the defined comfort requirements for the winter season—running mean ambient air temperature below 15 °C—are similar for all comfort models. Room temperatures are limited to a range of 19 to 24 °C.

On the contrary, comfort requirements differ significantly for the summer season, especially for the adaptive and static comfort models. Considering the latter, the defined setpoints for room temperature range between 20 and 26 °C (comfort class II). For the adaptive models, the room temperature setpoints increase with higher ambient air temperatures. Although thermal-comfort results differ significantly between European buildings, most of the buildings comply with the temperature requirements of the adaptive models. This does not hold true for the static comfort models, where temperature limits are often violated.

Figure 3.2 presents the thermal footprint of the European buildings: thermal comfort is evaluated for all six comfort models, considering the upper temperature limits during the summer season. Obviously, different comfort results are achieved, depending on the comfort model applied.

In summary, comfort models defined in various standards and guidelines result in different conclusions. In some cases, results differ considerably, in particular for the adaptive versus the static approach. Therefore, it is important to develop a classification of cooling concepts and the corresponding thermal-comfort models.

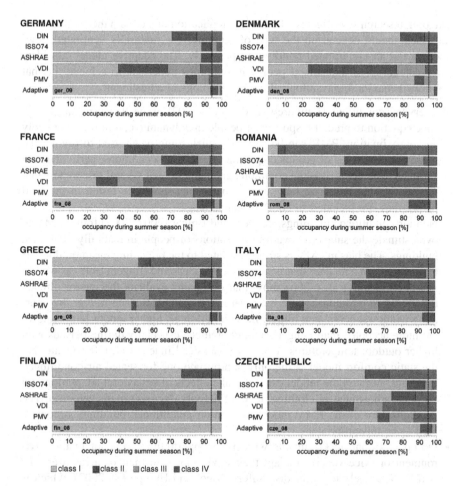

Fig. 3.2 Thermal comfort of eight European buildings (for description of the buildings, see Sect. 6.1), evaluated in accordance with the introduced standards. Abbreviations as follows: DIN 1946 (DIN), ASHRAE 55 (ASHRAE), German annex to DIN EN 15251-NA (VDI), PMV approach in EN 15251:2007-08 (2007) (PMV), adaptive approach in EN 15251:2008-07 (Adaptive). Thermal comfort with respect to class II should be guaranteed during 95 % of the occupancy (dotted line). Example: for the Danish building, thermal comfort is in compliance with class II during 95 % of the occupancy when considering the adaptive comfort models and with class III when considering the PMV comfort model. However, the operative room temperatures are higher than permitted during the entire summer season for the VDI and DIN 1946 models. Squares indicating classes I to IV are ordered from left to right

3.3 Discrepancy Between Static and Adaptive Approaches to Thermal Comfort

The last few decades of thermal-comfort research have produced an irreconcilable debate between two philosophies: the "PMV" and the "adaptive" model have contrasting assumptions about the way people respond to their environment.

Ongoing research investigates whether the application of a certain thermal-comfort approach should depend on the type of building, e.g., naturally ventilated, low-energy, and air-conditioned building, respectively. The adaptive model considers three categories of adjustment people in buildings undertake to achieve thermal comfort: behavioral, physiological, and psychological adjustment.

- Everyday condition: Humphreys and Nicol (2002) argue that using a steady-state equation to predict responses of people in a dynamic equilibrium is merely an approximation. PMV can be seriously misleading when used to predict the mean comfort votes of groups of people in everyday conditions, particularly in warm environments. It is the common agreement among several researchers that the relationship based on laboratory experiments should be tested in the field before they are included in comfort standards.
- Non-air-conditioned buildings: de Dear and Brager (2001, 2002) found PMV to overestimate the subjective warmth sensations of people in naturally ventilated buildings. The bias in PMV is strongly related to the prevailing mean outdoor air temperature, in a manner that is nonlinear. De Dear and Brager (1998) as well as van der Linden et al. (2006) show that occupants evaluate the indoor climate differently in buildings where they can individually influence it, e.g., by oper-ating windows, doors, and blinds. In such buildings, higher indoor temperatures are more accepted than Fanger's model predicts, especially during periods with higher outdoor temperatures. Therefore, van der Linden et al. (2006) state that the static comfort model is the only appropriate one for sealed air-conditioned buildings, because the model can only take effects of behavioral adaptation into account (adjustment of clothing, level of activity, increase in air velocity). The generalization of the PMV model for non-air-conditioned buildings is consid-ered to be inappropriate.
- Occupant's control: Occupants with more opportunities to adapt to the envi-ronment or (vice versa) to adapt their environment to their own requirements will be less likely to suffer discomfort (Nicol and Humphreys 2002), which is increased if control is not provided, or if the controls are ineffective, inappro-priate, or unusable. Bordass et al. (2001a, b) and Gossauer (2008) confirm these findings through comprehensive occupant surveys in the UK and Germany, respectively. Raja et al. (2001) as well as Rijal et al. (2007) state that windows are extensively used by occupants. At indoor temperatures above 20 °C, the number of subjects reporting the opening of windows increases strongly with indoor temperature and approaches 100 % at more than 27 °C. The importance of individual occupant control is already acknowledged by the German orga-nization 'Deutsche Gesellschaft für Nachhaltiges Bauen' (DGNB 2009), where occupants' influence on the surrounding condition is one factor for building certification. In addition, the U.S. 'Green Building Council's Leadership in Energy and Environmental Design' accreditation system has three aspects associated with thermal comfort, out of which one specifically applies to the control of operable windows (U.S. Green Building Council, LEED 2006).

- Occupants' expectation: Thermal sensation is influenced by an individual's experience and expectation of the building's climate, based on the outdoor temperature of that particular day and the preceding ones. However, results of a study by Gossauer (2008) in Germany reveal that occupants might appreciate the opportunity to condition the office or to influence the temperature at the workplace, respectively, at higher ambient air temperatures.
- Effectiveness of intervention: Post-occupancy evaluation studies demonstrate that occupant satisfaction with comfort correlates with opportunities to make interventions and with their effectiveness (Gossauer 2008; Leaman and Bordass 2001). Simplicity and convenience of intervention are paramount.
- Overall comfort: Intensive research based on field studies has determined that satisfaction with workplace conditions is not exclusively driven by thermal comfort, but is also affected by very diverse factors such as visual and acoustic comfort, interior design, the occupant's expectations and control of the surroundings, the user's behavior as well as social, cultural, and psychological parameters (Wagner et al. 2007; Leaman and Bordass 2001, 2007; Bischof et al. 2003; Humphreys and Nicol 2000a; Humphreys and Hancock 2007; Hellwig 2005). It is often stated that physiological acclimatization to heat or cold does not affect the preferred bodily condition to thermal comfort, but it is generally agreed that it does affect people's tolerance to body states differing from it (Humphreys and Nicol 2002).
- Health: Roulet (2001) states that, obviously, excessive energy consumption by the HVAC system does not result in better health. As a general rule, the higher the energy consumption, the larger is the number of sick-building syndromes (generally found in fully air-conditioned buildings (Bischof et al. 2003)).
- Forgiveness: In passively air-conditioned buildings (openable windows, sunshading systems), more adaptive mechanisms are typically available to the occupant for comfort and consequently support a greater individual awareness of the available adaptive opportunities. Buildings in which people have easy access to a variety of building controls enabling direct effects on comfort are found to show an attitudinal shift to occupant "forgiveness." Leaman and Bordass (2007) coined this term as a description of how people extend their comfort zone by overlooking and allowing for inadequacies of the thermal environment (Kwok and Rajkovich 2010). Forgiveness factors are defined for different building types, sorted in accordance with the ventilation system employed. Buildings with natural ventilation are found to have the highest forgiveness scores, i.e., people may be more likely to tolerate otherwise uncomfortable conditions in buildings with natural ventilation.
- Occupant's tolerance: Investigations by Humphreys and Nicol (2000b) show that it is increasingly common for users to tolerate hotter conditions than a generation ago. On average, people were comfortable at about 20 °C in 1978 (from data collected earlier) and at about 23 °C in 1998 (from data compiled by de Dear and Brager in 1998). Low-energy buildings tend to be warmer in summer than conventional air-conditioned buildings, but thermal comfort in them has been rated as more comfortable and satisfying (Leaman and Bordass 2007). However, it needs

to be verified if the 1998 findings are applicable to today's comfort requirements, since people spend a lot of time in air-conditioned spaces such as shopping malls, cars, trains, banks, etc. Therefore, higher indoor temperatures might now be perceived as more unpleasant and uncomfortable.

- Dress code: Changes in clothing and activity also change the conditions under which people feel comfortable. If no dress code is required, the occupant can adapt more easily to the surrounding conditions by changing clothes (Nicol and Humphreys 2002) and might therefore accept higher indoor air temperatures. Occupants are more satisfied with room temperatures if they do not have to adhere to a dress code (Gossauer 2008). Haldi and Robinson (2008) found by means of questionnaires that occupants are adaptive in terms of clothing if they do not have to adhere to a formal dress code. They found a clear linear relationship between the running mean ambient air temperature and the level of clothing insulation. Outdoor temperature was found to be twice as effective as indoor temperature to explain the variation in clothing level.
- Productivity: The great challenge in correlating interior comfort and occupant productivity is that there is no common agreement on the definition and the methodology to measure productivity in office work. Some researchers argue that studies on the relation between warmer environments and productivity are not very conclusive (van Linden et al. 2006). Some studies demonstrate a decrease in productivity at higher ambient temperatures [e.g., by Wyon qt. in (Fitzner 2004)]. Others show a positive effect on productivity when people have adaptive opportunities, like openable windows, fans and blinds to alter their subjective warmth. In addition, field studies reveal that perceived productivity does not vary with indoor air temperature and that productivity positively correlates with the perception of general comfort and health. Therefore, some researchers believe that adaptive comfort models in moderate climates will have no adverse affect on productivity, provided that adaptive opportunities are available. Occupants who perceive that they feel comfortable, also tend to say they feel healthy and productive at work. Thus, health, comfort, and productivity are often surrogates to each other (Bordass et al. 2001a). Fitzner (2003) presents a literature review on productivity.
- Energy Savings: On the one hand, the provision of occupant comfort has a major bearing on energy consumption. On the other hand, an energy-efficient building that cannot provide comfortable and high-quality working conditions will either affect the well-being—and therefore the productivity—of the occupants or drive them into taking actions that may compromise the energy economy of the building (e.g., subsequent installation of portable cooling devices) (Nicol 2007). In brief, an energy-efficient concept without occupant comfort compromises the sustainability and profitability of the property. As a result, an optimum is needed between energy efficiency, interior comfort, and expenditure, both for new constructions and refurbishments. The EPBD (Energy Performance of Buildings Directive) Article 7 requires the inclusion of information on the interior climate of a building for the certification of energy use. Besides, post-occupancy evaluation and field studies show that high energy use for the heating and

cooling of buildings does not necessarily correlate with high occupant satisfaction. The sustainability of building and technical plant performance needs to be considered in the framing of standards (Roulet et al. 2006a, b; Humphreys and Nicol 2002; Nicol and Humphreys 2009).

References

ASHRAE Standard 55-2004 (2004) Thermal Environmental Conditions for Human Occupancy. American Society of Heating, Refrigerating and Air-Conditioning Engineers Inc., Atlanta

BBR-Lex (2008) Richtlinie B12–8132.1/0 zu baulichen und planerischen Vorgaben für Baumaßnahmen des Bundes zur Gewährleistung der thermischen Behaglichkeit im Sommer. Bundesministerium für Verkehr, Bauen und Stadtentwicklung, Berlin

Bischof W, Bullinger-Nabe M, Kruppa B, Schwab R, Müller BH (2003) Expositionen und gesundheitliche Beeinträchtigungen in Bürogebäuden—Ergebnisse des ProKlimA-Projektes. Fraunhofer IRB Verlag, Stuttgart

Boersta AC, Hulsman LP, van Weele AM (2005) ISSO 74 Kleintje Binnenklimaat. Stichting ISSO, Rotterdam

Bordass B, Cohen R, Standeven M, Leaman A (2001a) Assessing building performance in use 2: technical performance of the probe buildings. Build Res Inf 29(2):103–113

Bordass B, Cohen R, Standeven M, Leaman A (2001b) Assessing building performance in use 3: Energy performance of the probe buildings. Build Res Inf 29(2):114–128

CEN/CR 1752 (2001) Ventilation for buildings—Design criteria for the indoor environment. European Committee for Standardization, Brussels

de Dear R (1998) A global database of thermal comfort field experiments. ASHRAE Trans 104(1b):1141–1152. Reprinted in Schiller Brager G (ed) (1998) Field studies of thermal comfort and adaptation. ASHRAE Techn Data Bull 14(1):15–26

de Dear R, Schiller Brager G (1998) Developing an adaptive model of thermal comfort and preference. ASHRAE Trans 104(1a):pp.145–167. Reprinted in Schiller Brager G (ed) (1998) Field studies of thermal comfort and adaptation. ASHRAE Techn Data Bull 14(1):27–49

de Dear R, Brager G (2001) The adaptive model of thermal comfort and energy conservation in the built environment. Int J Biometeorol 45:2s

de Dear R, Brager G (2002) Thermal comfort in naturally ventilated buildings—revisions to ASHRAE Standard 55. Energy Build 34(6):549–561

Deutsche Gesellschaft für Nachhaltiges Bauen e. V. (DGNB) (2009) http://www.dgnb.de. Accessed Dec 2013

DIN 1946:1994-01 (1994) Ventilation and air-conditioning: technical health requirements. Beuth Verlag, Berlin

DIN EN 13779:2004-09 (2004) Ventilation for non-residential buildings—Performance requirements for ventilation and room-conditioning systems. Beuth Verlag, Berlin

EN 15251:2008-07 (2007) Criteria for the indoor environment. Beuth Verlag, Berlin

Fanger PO (1970) Thermal comfort analysis and applications. Environmental Engineering. Mc-Graw-Hill, New York

Fitzner K (2003) Productivity. In: Nilsson PE (ed) Achieving the desired indoor climate. Studentlitteratur AB, Stockholm

Fitzner K (2004) Einfluss des Raumklimas auf die Produktivität. HLH 55(9):59–60

Gossauer E (2008) Nutzerzufriedenheit in Bürogebäuden—eine Feldstudie. Dissertation, University of Karlsruhe. Fraunhofer IRB Verlag, Stuttgart

Gossauer E, Wagner A (2008) Occupant satisfaction at workplaces—a Field study in office buildings. In: Proceedings of Windsor conference on air-conditioning and the low carbon cooling challenge, London Metropolitan University, Windsor, July 2008

Haldi F, Robinson D (2008) On the behavior and adaptation of office occupants. Build Environ 43(12):2163–2177

Hellwig R (2005) Thermische Behaglichkeit, Disserion. TU München

Humphreys MA, Hancock M (2007) Do people like to feel 'neutral'?: exploring the variation of the desired thermal sensation on the ASHRAE scale. Energy Build 39(7):867–874

Humphreys MA, Nicol JF (2000a) Effects of measurement and formulation error on thermal comfort indices in the ASHRAE database of field studies. ASHRAE Transactions 206:493–502

Humphreys MA, Nicol JF (2000b) Outdoor temperature and indoor thermal comfort: Raising the precision of the relationship for the 1998 ASHRAE database of field studies. ASHRAE Transactions 206:485–492

Humphreys MA, Nicol JF (2002) The validity of ISO-PMV for predicting comfort votes in every-day thermal environments. Energy Build 34(6):667–684

ISO EN 7730:2005 (2005) Ergonomics of the thermal environment—Analytical determination and interpretation of thermal comfort using calculation of the PMV and PPD indices and local thermal comfort criteria. International Organization for Standardization, Genova

Kwok AG, Rajkovich NB (2010) Addressing climate change in comfort standards. Build Environ 45(1):18–22

Leaman A, Bordass B (2001) Assessing building performance in use 4: the probe occupant surveys and their implications. Build Res Inf 29(2):129–143

Leaman A, Bordass B (2007) Are users more tolerant of 'green' buildings? Build Res Inf 35(6):662–673

Nicol JF (2007) Comfort and energy use in buildings—getting them right. Energy Build 39(7):737–739

Nicol JF, Humphreys MA (2002) Adaptive thermal comfort and sustainable thermal standards for buildings. Energy Build 34(6):563–572

Nicol JF, Humphreys MA (2009) Derivation of the adaptive equations for thermal comfort in free-running buildings in European standard EN15251. Build Environ 45(1):11–17

Nicol JF, McCartney K (2002) Developing an adaptive control algorithm for Europe. Energy Build 34(6):623–635

Nicol JF, Roaf S (2005) Post-occupancy evaluation and field studies of thermal comfort. Build Res Inf 33(4):338–346

Raja IA, Nicol JF, McCartney KJ, Humphreys MA (2001) Thermal comfort: use of controls in naturally ventilated buildings. Energy Build 33(3):235–244

Rijal HB, Tuohy P, Humphreys M, Nicol F, Samuel A, Clarke J (2007) Using results from field surveys to predict the effect of open windows on thermal comfort and energy use in buildings. Energy Build 39(7):823–836

Roulet CA (2001) Indoor environment quality in buildings and its impact on outdoor environment. Energy Build 33(3):183–191

Roulet CA, Johner N, Foradini F, Bluyssen P, Cox C, Fernandes E, Müller B, Aizlewood C (2006a) Perceived health and comfort in relation to energy use and building characteristics. Build Res Inf 34(5):467–474

Roulet CA, Flourentzou F, Foradini F, Bluyssen P, Cox C, Aizlewood C (2006b) Multicriteria analysis of health, comfort and energy efficiency in buildings. Build Res Inf 34(5):475–482

U.S. Green Building Council, LEED (2006) New construction reference guide, 2nd edn. USGBC, Washington

van der Linden AC, Boersta AC, Raue AK, Kuvers SR, de Dear RJ (2006) Adaptive temperature limits: a new guideline in The Netherlands—a new approach for the assessment of building performance with respect to thermal indoor climate. Energy Build 38(1):8–17

Wagner A, Gossauer E, Moosmann C, Gropp T, Leonardt R (2007) Thermal comfort and workplace occupant satisfaction—results of field studies in German low-energy office buildings. Energy Build 39(7):758–769

Chapter 4
User Satisfaction with Thermal Comfort in Office Buildings

Abstract Thermal comfort in nonresidential buildings is evaluated in accordance with the European standard EN 15251 which defines two comfort models based on the cooling concept implemented in the building: the adaptive model and the PMV model. However, many office and administration buildings cannot be clearly allocated to a specific comfort model according to the respective cooling concept employed. A field study confirms that the standard for thermal interior comfort should provide two models. However, the strict allocation of building categories (building with/without mechanical cooling) in EN 15251 could not be verified. It seems reasonable to classify buildings into air-based and water-based low-energy cooling systems with limited cooling capacity or to do so with respect to the degree of coupling between indoor and outdoor climate conditions.

Thermal comfort in nonresidential buildings is evaluated in accordance with the European standard EN 15251:2007–08 (2007) which defines two comfort models based on the cooling concept implemented in the building: the adaptive model and the PMV model (Chap. 3). However, many office and administration buildings in Europe cannot be clearly allocated to a specific comfort model according to the respective cooling concept employed. For example, some buildings use free cooling or mechanical night-ventilation in combination with active cooling devices (e.g., thermo-active building systems using water as the active medium). Furthermore, cooling concepts using environmental heat sinks (e.g., direct cooling by means of geothermal energy) constitute energy-efficient solutions for Central European climates, but cannot ensure stringent room-temperature setpoints with respect to the PMV comfort model due to system inertia and system-related temperature differences.

Since the standard must formulate unambiguous conditions, DIN 15251 contains the following definition:

> Buildings without mechanical cooling: buildings that do not have any mechanical cooling and rely on other techniques to reduce high indoor temperature during the warm season like moderately sized windows, adequate sun shielding, use of building mass, natural ventilation, night-ventilation, etc. to prevent overheating.

D. E. Kalz and J. Pfafferott, *Thermal Comfort and Energy-Efficient Cooling of Nonresidential Buildings*, SpringerBriefs in Applied Sciences and Technology, DOI: 10.1007/978-3-319-04582-5_4, © The Author(s) 2014

In this context, "mechanical cooling" is defined explicitly and is distinguished from passive cooling methods in terms of the guideline as follows:

Cooling of the indoor environment by mechanical means used to provide cooling of supply air, fan-coil units, cooled surfaces, etc. The definition is related to people's expectations regarding the internal temperature in warm seasons. Opening of windows during day and night time is not regarded as mechanical cooling. Any mechanically assisted ventilation (fans) is regarded as mechanical cooling.

The term "mechanically cooled" encompasses all concepts employing a mechanical device to condition the space, such as supply and/or exhaust air systems, thermo-active building systems, and convectors. Only buildings employing natural ventilation through open windows fall into the category of "non-mechanical" concepts. This method may be applied when certain requirements are met: thermal conditions are primarily regulated by the occupants by operating windows that open to the outdoors. Furthermore, occupants are engaged nearby in sedentary activities and are supposed to feel free to adapt their clothing to thermal conditions.

This strict definition of the standard EN 15251 according to the cooling concept employed, however, differs significantly from the conditions formulated in the adaptive comfort model, which is guided by the possibility of effective user influence and not by the cooling concept, e.g., (van der Linden et al. 2006) and (ISSO 74 2004). Which definition is better to describe the comfort model to be used in nonresidential buildings with technically limited cooling capacity: the technical cooling equipment or the users' expectations? Several publications refer to this unanswered question (de Dear and Brager 1998; Humphreys and Nicol 2000), but a definitive proposal fails due to the small sample sizes of such buildings in field studies (de Dear and Brager 2002).

The objective is to assess the occupant perception of and satisfaction with thermal comfort in two office buildings located in close proximity to each other in Freiburg in south-west Germany, where two different cooling concepts are employed. Results were obtained in long-term field surveys through daily questioning and accompanying measurements in high temporal resolution. A model to predict the comfort temperature in summer was calculated by means of regression analyses, based on the available data from the field survey and compared to the models in the EN 15251 guideline (Kalz et al. 2013).

The study is designed to answer the following questions:

(1) Can an adaptive comfort model be used to assess thermal comfort in non-residential buildings? Do their occupants tolerate higher operative room temperatures with rising outdoor temperatures, provided that the occupants can influence the indoor environment?
(2) Do occupants have an altered perception of the operative room temperature subject to the cooling concept employed?
(3) Does user satisfaction with thermal comfort change depending on the cooling concept?

(4) How is the comfort temperature determined for different cooling concepts, considering the relationship between room temperature, ambient air temperature, and occupant satisfaction? How can cooling concepts be allocated to the defined comfort models in the standards? Should the standard be extended or complemented?

4.1 Methodology and Analysis of the Field Survey

Investigation and Participants. The study was conducted in two office buildings in Freiburg, located less than 500 m apart (Fig. 4.1). One of them (SIC) is air-conditioned in summer in accordance with a mechanical night-ventilation concept (exhaust ventilation system, "air-based"). The second building (SCF) employs thermo-active building systems with water as the active medium (concrete-core conditioning, "water-based") and a supply-and-exhaust air system (see Sect. 4.2). In both buildings, the occupants can influence the indoor environment, i.e., they can individually open/close windows, control the external solar-shading system, and a dress code is not observed.

The investigation was carried out during the summer of 2009 (July 1 to September 30) in the office rooms of the SIC building and in the following year, 2010 (June 1 to September 30), in the SCF office building. User perception of and satisfaction with thermal comfort was determined by means of daily, computer-based questionnaires.

In the SIC building (2009, night-ventilation concepts), 26 occupants participated in the survey; in the SFC building, there were 29 participants (Table 4.1). Some of the occupants interviewed in the SIC building moved to the SCF building in spring 2010. Therefore, twelve of them took part in both surveys.

Survey of User Perception and Satisfaction. Daily computer-based surveys were carried out during the period of investigation. A query appeared on the users' computer screens at 11 a.m. and 3 p.m., in which the occupants were asked to answer four questions about thermal room comfort. The questions and the scaling were as follows:

- How do you perceive the operative room temperature to be at the moment? (with 7 = hot; 1 = cold)
- How satisfied are you with the operative room temperature at the moment? (with 4 = very dissatisfied; 1 = very satisfied)
- Do you perceive any air movement/air draft at the moment? (with 5 = not at all; 1 = very much)
- How do you perceive the air humidity to be at the moment? (with 5 = very dry; 1 = very humid)

Table 4.1 Information on method and execution of postoccupancy evaluation as well as measurements

	Air-based cooling (SIC)	Water-based cooling (SCF)
Participants		
Number	26 (9 female and 17 male occupants)	29 (9 female and 20 male occupants)
Age	Up to 25 years: 4 users	Up to 25 years: 2 users
	Between 26 and 35 years: 18 users	Between 26 and 35 years: 19 users
	Between 36 and 45 years: 3 users	Between 36 and 45 years: 8 users
	Between 46 and 55 years: no users	Between 46 and 55 years: no users
	Over 55 years: 1 user	Over 55 years: no user
Questioning		
Period	07/01/09–09/30/09	06/01/10–09/30/10
Frequency	Daily; in the morning at 11 a.m. and in the afternoon at 3 p.m.	
	Note: The question stays on the screen until the user answers the question. Therefore, the response times vary slightly if a person is not at the workplace at the time of questioning	
Methodology	Computer-based questioning	
Questions	1. How do you perceive the operative room temperature to be at the moment? (with 7 = hot; 1 = cold)	
	2. How satisfied are you with the operative room temperature at the moment? (with 4 = very dissatisfied; 1 = very satisfied)	
	3. Do you feel any air movement/air draft at the moment? (with 5 = not at all; 1 = very much)	
	4. How do you perceive the air humidity to be at the moment? (with 5 = very dry; 1 = very humid)	
Measurements		
Location	At the workplace of the occupants interviewed	
Indoor climate	Operative room temperature [°C] and relative air humidity (%) in 12-minute resolution	
User behavior	Opening of windows, status of solar shading	No measurements
Number of rooms	10	14
Outdoor climate	Ambient air temperature [°C], solar radiation [W/m^2], wind velocity [m/s], ambient air humidity [%] in 1-min resolution at local meteorological station	
Possible Influential Factors		
	No dress code	
	Manual operation of solar shading	
	Manual operation of windows	
	Minimum 1 window per 2 occupants	
	No influence on setpoints of operative room temperature due to night-ventilation concept	Setpoint control in 2-Kelvin steps

The users' assessment of the indoor temperature conditions was rated on a 7-point scale, while air movement and humidity were rated on a 5-point scale. For both scales, only the endpoints are named explicitly, not the intermediate gradations. The scale for the question of user satisfaction was evenly numbered in order to avoid neutral responses, and thus to obtain a clear distinction between "satisfied" and "dissatisfied."

Survey of User Behavior. In addition to the daily questioning, occupants of the SIC building were asked on four summer days, with high ambient air temperatures which measures they had taken each morning and afternoon respectively to improve their (perceived) thermal comfort. The questioning was conducted by means of a detailed, written questionnaire. Besides, the occupants were asked to self-assess their productivity at the time of questioning under the prevailing environmental indoor conditions.

Monitoring of Indoor and Outdoor Conditions. Parallel to the questioning, the following parameters to quantify thermal comfort were monitored in high temporal resolution (time step: 12 min) in the rooms or at the workplaces respectively:

- Measurements at the workplace of the occupants: operative room temperature and relative air humidity.
- Measurements in the offices: opening of windows and status of the solar-shading devices.
- Ambient conditions at local meteorological station: ambient air temperature, solar radiation, relative air humidity, wind velocity, and wind direction (Fig. 4.2)

Analysis. The computer-based data from the questionnaires and the workplace-related measurements at the time of questioning were combined to a total dataset and analyzed with SPSS (SPSS 20.0 2008). Regression analyses were used to model, first the relationship between outdoor and indoor air temperature as well as satisfaction, and second to determine the comfort room temperature.

4.2 Building and Energy Concepts of Demonstration Buildings

The demonstration buildings used for the field study strive for a significantly reduced cooling energy demand (around 25 kWh_{therm}/m^2a) with carefully coordinated measures for passive cooling and the use of environmental heat sinks (ambient air and groundwater). Following a stringent load-reduction strategy, limited use of primary and final energy was set as a target for the complete service technology of the building (HVAC and lighting). The measures included a high-quality building envelope, reduced solar heat gains (through solar shading), sufficient thermal storage capacity (through nonsuspended concrete ceilings), air-tight building envelope in conjunction with a hygienically compulsory air-ventilation

Fig. 4.1 Office buildings: SIC (*left*) with mechanical night-ventilation concept and SCF (*right*) with water-based cooling concept. (Research for Energy Optimized Buildings (EnOB), www. enob.info/eng/ and UNMÜSSIG GmbH)

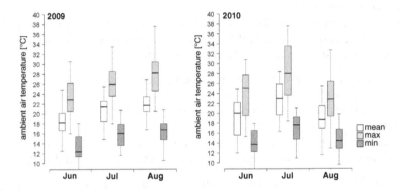

Fig. 4.2 Monthly ambient air temperature [°C] in summer, 2009 and 2010, location Freiburg: daily mean, maximum, and minimum. The rectangles present the data between the first and third quartiles. The maximum and minimum values are also plotted

system, and low-energy office equipment (reduced internal heat gains, daylight concepts) (see Table 4.2). Both buildings allow the user to influence the indoor environment with devices such as operable windows and sun-shading controls. In the SCF building, the user can control the setpoints for the operative room temperature in 2-Kelvin steps. Most of the office buildings consist of 2- and 4-person as well as group offices. The energy concept of both buildings for heating, cooling, and ventilation is illustrated in Fig. 4.3.

Ventilation concept: During the time of occupancy, the SIC building is manually ventilated via windows. The SCF building employs a supply-and-exhaust ventilation system, which enables the supply air to be cooled down slightly by a water-to-air heat exchanger that is coupled to a groundwater system.

Cooling concept: In summer, the SIC building is cooled down with a night-ventilation concept by using an exhaust ventilation system (1.5 air changes per hour). Night-ventilation is controlled through comparison between the average room and the outside temperature, and is activated only during summer. Outside

Table 4.2 Information on the SIC and SCF buildings

	SCF	SIC
Building and usage		
Usage, number of users	Office, -	Office, 400
Time of occupancy	8:00–20:00	6:00–20:00
	Mon–Fri	Mon–Fri
Year of completion	2010	2003
Number of storeys	5	5
Heated net floor area [m^2]	4,500	13,833
Area-to-volume ratio [m^{-1}]	0.33	0.29
Building envelope		
Solar-shading system	Exterior venetian blinds, central control, shading factor 0.2	Exterior venetian blinds, central control, shading factor 0.2
U value [W/(m^{2}K)]	No information	Exterior walls 0.19
		Windows 1.3
		Roof 0.19
Windows	Double-glazed, low-e windows	Double-glazed, low-e windows g-value 0.60 window-to-wall ratio 33–49 %
Cooling concept		
Environmental heat sink	Ambient air, groundwater	Ambient air
Input energy form	Electricity	Electricity
Cooling system	Direct cooling via groundwater system and wet cooling tower, additional compression chiller and cooling tower	Mechanical night-ventilation
Capacity [kW$_{therm}$]	227	–
Distribution system in the room	Thermo-active building systems	Air
Ventilation concept		
Operable windows	Yes	Yes
Night-ventilation	No	Yes
Mechanical ventilation	Yes	Yes
Dehumidification of air	No	No
Pre-cooling	Yes	No

air flows into the room through the air inlets that are integrated into the upper window frame and is exhausted through an aperture in the duct system.

The SCF building is cooled down actively by thermo-active building systems (here: concrete-core conditioning system, CCT), where pipes are integrated into the core of the concrete ceiling of the office rooms (20 mm pipe diameter, 150 mm spacing between pipes). The system is designed for a supply temperature of 16 °C and a temperature difference of 3 K. Each office has an individual room temperature control (setpoint controller in 2-Kelvin increments), resulting in 263

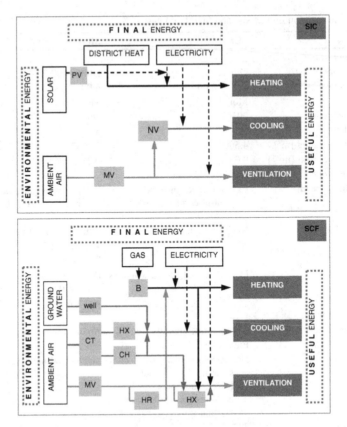

Fig. 4.3 Energy scheme of the SIC and SCF buildings. Abbreviations mechanical ventilation (MV), night-ventilation (NV), photovoltaics (PV), gas boiler (B), heat recovery (HR), heat exchanger (HX), cooling tower (CT), chiller (CH)

hydronic circuits in the entire building, which are controlled by electronic actuators.

Cooling energy is generated either through direct use of a groundwater system (temperature level in summer between 14 and 19 °C, capacity about 200 kW$_{therm}$) or by a wet cooling tower (200 kW$_{therm}$). If demand is higher, cooling energy can also be provided by two compression chillers.

4.3 Results of Survey and User Satisfaction

In the daily survey, users rated their perception of and their satisfaction with the room temperature at the time of questioning (7-point scale for perceptions from "cold" to "hot" and 4-point scale for satisfaction from "very dissatisfied" to

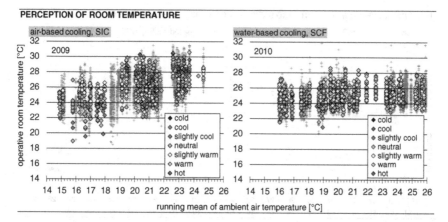

Fig. 4.4 Perception of room temperature by the users: hourly operative room temperature [°C] at all workplaces of the users interviewed plotted against the running mean of the daily ambient air temperature [°C]. Also given is user perception of the room temperature, following a 7-point scale from "cold" to "hot." Note: *colored markers* present the measurements at the time of questioning; *gray markers* present all measurements during occupancy

"very satisfied"). Results for perceived and actually measured room temperature are shown in Figs. 4.4 and 4.5. Figure 4.4 shows the hourly measured room temperature during summer period (gray markers) as well as the its perception at the time of questioning (colored markers). The surveys in both buildings indicate that room temperatures above 26 °C are usually perceived as "slightly warm" to "warm." Room temperatures above 28 °C were rated as "warm" to "hot." Measured room temperatures down to 20 °C were never perceived as "cold." Figure 4.5 relates the perception of the room temperatures to user satisfaction:

- Room temperatures below 22 °C were perceived as "slightly cool."
- Considering both cooling concepts, satisfied users generally rate room temperatures up to 25 °C as "neutral," from 26 to 28 °C as "slightly warm" and beyond that as "warm."
- Unexpectedly, there were quite a few dissatisfied users for room temperatures up to 26 °C: 11 % in the SIC building and 8 % in the SCF building. Thereby, temperature perception is scattered across a wide range from "cool" to "warm."
- For both cooling concepts, dissatisfied users rated temperatures above 26 °C as "warm."

4.4 Satisfaction with Room Temperature

Figure 4.6 shows the degree of user satisfaction with the room temperature. The ratings "very satisfied" and "satisfied" are combined, as are "very dissatisfied" and "dissatisfied." Considering the overall satisfaction with the room

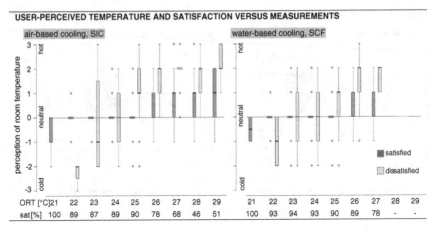

Fig. 4.5 Users' perception of room temperature compared to measurements of and satisfaction with room temperature: perception of room temperature by users on a 7-point scale (from "cold" to "hot") at a given operative room temperature (x-axis) at the time of questioning. Results are color-coded according to the degree of satisfaction, i.e., users are "very satisfied"/"satisfied" (*dark gray*) and "dissatisfied"/"very dissatisfied" (*light gray*) with prevailing room temperature. Note: measured operative room temperature (ORT) is categorized into < 21, 21–22, 22–23, 23–24, 24–25, 26–27, 28, 29, > 29 °C. Additionally, percentage of users satisfied (sat) with the room temperature [%] in the temperature classes is shown

temperatures, there are evident differences between the cooling concepts. Figure 4.6 illustrates the hourly room temperatures plotted against the running mean ambient air temperature, with an indication of user satisfaction:

- In the SIC building (night-ventilation), 77 % of the users were satisfied with the room temperature over the entire period of the survey. By contrast, in the SCF building (water-based cooling), 91 % of the interviewed occupants were satisfied with the room temperature.
- The number of occupants dissatisfied with the room temperature in the SIC building correlated with the room temperature, i.e., the number of dissatisfied occupants increased with rising room temperatures. Obviously, a room temperature of 26 °C is a pronounced threshold for dissatisfaction with thermal indoor comfort. Regardless of the prevailing outdoor temperatures, 89 % of the users were satisfied with room temperatures below 26 °C. Even 74 % of them were still satisfied with room temperatures ranging between 26 and 28 °C. Still, higher room temperatures lead to a significant proportion of dissatisfied users, namely 52 %.
- In the SCF building, room temperatures with a maximum of 28 °C were recorded for only relatively few hours. This does not allow for a direct comparison with the SIC building; here, the maximum room temperatures exceeded 29 °C. However, the dissatisfaction of the users in the SCF building seems to have increased significantly at room temperatures above 26 °C.

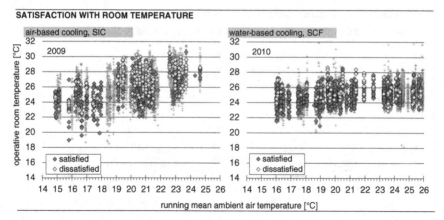

Fig. 4.6 Occupant satisfaction with room temperature: Hourly measured operative room temperature [°C] plotted against the running mean ambient air temperature [°C] during time of occupancy: room temperature measurements (*light gray*), user is "very satisfied" or "satisfied" with the room temperature (*dark gray*) and user is "dissatisfied" or "very dissatisfied" with the room temperature (*white*)

4.5 Determination of the Comfort Temperature

Based on the data compiled during the survey, a model for predicting the comfort temperature was developed by means of a regression analysis. After several test runs, the combination of running mean ambient temperature (calculation in accordance with EN 15251:2007–08 (2007)), actual room temperature at the time of questioning, and the question"How satisfied are you with the room temperature?" delivered the most reliable correlation.

The objective was to determine the model that best predicts when the number of occupants satisfied with a given room temperature is highest as a function of the ambient air temperature. Therefore, a function for the room temperature in dependence of the running mean ambient air temperature was first calculated by regression only for the case of users being satisfied with the prevailing thermal room conditions. This means that the indoor conditions were determined under which the users were satisfied. The calculated regression equation was set off against a regression equation relating the room temperature to the running mean ambient air temperature for the case that users were dissatisfied with the room temperature.

This resulted in a Pareto-optimal prediction model which predicts the conditions, where the number of satisfied users is at its maximum and the number of dissatisfied users at its minimum. Results are given in Fig. 4.7.

Central results for the SIC building with air-based cooling are:

- The indoor comfort temperature is determined to be $RT_c = 17.78\ °C + 0.34\theta_{e,rm}$. Based on the available data, the comfort model could be developed for a daily mean ambient air temperature of 16 °C and above.

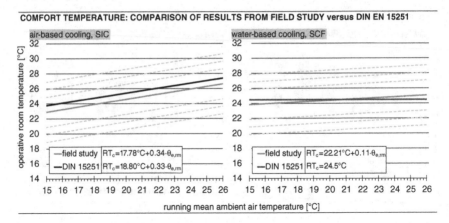

COMFORT TEMPERATURE: COMPARISON OF RESULTS FROM FIELD STUDY versus DIN EN 15251

Fig. 4.7 Comparison of comfort temperature (RT$_c$) derived from occupancy evaluation and guideline EN 15251

- The comfort boundaries I to III of 2, 3 and 4 K, defined by the EN 15251 standard, result in fractions of satisfied users of 85, 81, and 77 % respectively.
- A significant linear impact of the running mean ambient air temperature can be observed in the derived equation for the comfort temperature. The warmer the outside temperature, the higher is the room temperature at which the users are still satisfied.
- Consequently, the running mean ambient air temperature provides a reasonable reference value for assessing interior thermal comfort.
- The analysis of the survey and the resulting determined comfort temperature, both confirm the adaptive comfort model of EN 15251:2007–08 (2007). The equations for the comfort temperature are almost identical: user survey $RT_c = 17.78\ °C + 0.34\theta_{e,rm}$ and standard EN 15251 $RT_c = 18.8\ °C + 0.33\theta_{e,rm}$. For the equation derived from the user survey, there is only a minor correction of 0.8 K toward lower room temperatures.
- Based on the results of the field study, the application of an adaptive comfort model for the evaluation of thermal comfort in nonresidential buildings with a night-ventilation concept appears to be suitable. According to this model, occupants tolerate higher room temperatures if the outside temperature is higher. However, user satisfaction is expected to decline significantly for room temperatures exceeding 27 °C. Therefore, the definition of an upper limit for the maximum permissible room temperature appears to be relevant.
- The violation of the defined lower comfort boundaries (21–24 °C) appears to be noncritical. Room temperatures of 21–24 °C are perceived to be"slightly cool" to "neutral." User satisfaction is at around 89 %. However, temperatures below the lower comfort boundaries can be avoided through a system control based on outside and room temperatures. In addition, the required thermal cooling energy is reduced.

- The study shows that occupants adapt to the interior and exterior conditions by taking measures, if possible, to counteract increasing temperatures during the day and thus positively influence their temperature perception. Measures taken by the users interviewed were documented on four hot summer days by means of a detailed questionnaire every morning and afternoon. Obviously, users tried to regulate the indoor climate mainly by opening windows or lowering the solar-shading devices. Individual pieces of clothing were not often taken off or put on. Overall, the users took action slightly more often in the morning than in the afternoon, although the room temperatures were higher in the afternoon.

Central results for the SCF building with water-based cooling:

- The indoor comfort temperature is determined to be $RT_c = 22.21\ °C + 0.11\theta_{e,rm}$. Based on the available data, the comfort model could be developed for a daily mean ambient air temperature of 16 °C and above.
- In comparison to the first field study, the outside temperature has little impact on determining the comfort temperature. Therefore, the ambient air temperature plays a minor role as a reference value. Based on the user survey, the comfort temperature can be calculated by using the same equation regardless of whether the current, the mean daily, or the running mean outdoor temperature is used as the reference.
- The calculated comfort temperature is almost equivalent to the PMV comfort model of EN 15251:2007–08 (2007). The comfort boundaries I to III of 1, 1.5 and 2 K, as defined by the standard, result in a fraction of satisfied users of 92, 91 and 90 % respectively.
- Temperatures lower than the defined lower comfort boundaries (22–24 °C) do not seem to be critical for water-based cooling concepts either. Room temperatures in this range are perceived to be "slightly cool" to "neutral," but user satisfaction is very high at about 92 %. However, temperatures below the lower comfort boundaries can be avoided through system controls based on outside and room temperatures. In addition, the required thermal cooling energy is reduced as well.
- The range of the ambient air temperature between 16 and 26 °C corresponds to a derived comfort temperature range of 24–25 °C. Hence, user acceptance of higher room temperatures at increasing ambient air temperatures does not seem to be applicable to buildings with water-based cooling concepts. The determination of room temperatures with more dissatisfied users, given the data obtained, is difficult, since the active water-based cooling of the offices prevents the increase of room temperatures above 27 °C. In addition, user satisfaction is above 90 %, even if the comfort temperature is not being considered at all.

The significant differences in user satisfaction with the room and the comfort temperature, depending on the cooling concept, are particularly noteworthy, considering that half of the users interviewed participated in both surveys, i.e., 12 out of 26 respondents. The same users rated their perception and satisfaction with the room temperature differently—according to the cooling concept employed in the building—, although the two buildings—are equipped almost identically—except for the cooling system. Various studies based on occupant surveys document that user satisfaction with room comfort is affected substantially by possibilities to exert influence on the room conditions and their effectiveness (e.g., Wagner et al. 2007). The present results, however, also imply that the expectation of users with room and comfort conditions has a decisive influence on their perception and satisfaction: users expect higher room temperatures in a building without cooling or with limited cooling capacities (such as in a night-ventilation concept) and even accept these temperatures, if they can adapt to the prevailing conditions. Correspondingly, users have higher expectations on the interior comfort in nonresidential buildings with water-based cooling concepts, and, therefore, are less satisfied with higher room temperatures.

Main results:

(1) The indoor climate in buildings with limited cooling capacities is rated positively by their users: 77 % of them were satisfied with the room temperatures in the building with the night-ventilation concept over the entire period of the study; in the building with the water-based cooling concept, there were 91 % satisfied users.

(2) The number of users dissatisfied with the room temperatures correlated with the room temperatures themselves, meaning that the number of dissatisfied users grows with rising room temperatures. Evidently, a room temperature of 26–27 °C is a distinct limit for satisfaction with thermal indoor comfort.

(3) The application of an adaptive comfort model for the evaluation of thermal comfort in nonresidential buildings with free or mechanical night-ventilation appears to be suitable. According to this model, occupants tolerate higher room temperatures at higher ambient air temperatures. However, user satisfaction is expected to decline significantly for room temperatures above 27 °C. The analysis of the survey and the resulting comfort temperature confirm the adaptive comfort model of EN 15251:2007–08 for buildings with night-ventilation.

(4) The determined comfort range for water-based cooling concepts with a higher cooling capacity is between 24 and 25 °C, which is very similar to the PMV comfort model of EN 15251:2007–08 (2007). Hence, the acceptance of users with higher room temperatures in correlation with higher ambient air temperatures can be applied only with restrictions to buildings with water-based cooling concepts.

(5) User satisfaction with thermal interior comfort is proven to be enhanced by the option to effectively influence room conditions. The present results also suggest that the expectations on room and comfort conditions have a considerable impact on perception and satisfaction: in buildings with night-ventilation, users expected higher room temperatures and thus accepted them. On the contrary, users had higher expectations on indoor comfort in buildings with water-based cooling concepts, and, therefore, were less satisfied with higher room temperatures.

The present study confirms that the standard for thermal interior comfort should provide two models. However, the strict allocation of building categories (building with/without mechanical cooling) in EN 15251 could not be verified. Similarly, the model of expectations, as used in the ISSO 74 standard, cannot be confirmed. Instead, it seems reasonable to classify buildings into air-based and water-based systems or to do so with respect to the degree of coupling between indoor and outdoor climate conditions.

In the first case, the classification is defined by the capacity and limits of the cooling power of the technical system, in the second case by building physics and the development of the room temperature. In both cases, the relevant evaluation parameters are available from the design and planning processes.

As a result of the field study, the authors recommend a revision of the comfort standard. This does not primarily concern the adjustment of the comfort temperature or the comfort boundaries, but the allocation of cooling concepts to the particular comfort model. We recommend evaluating buildings with free and mechanical night-ventilation as well as with passive cooling measures (e.g., solar-shading devices, daytime ventilation) in accordance with the adaptive comfort model. By contrast, the indoor comfort in buildings should be evaluated in accordance with the PMV model, if air-conditioning systems, fan-coil cooling, or water-based radiant cooling systems are employed. Even in these buildings, users seem to adapt to the prevailing outdoor climate conditions. However, the users have especially high expectations on the interior thermal comfort and, therefore, tolerate only slightly higher room temperatures compared to the defined temperature setpoints in EN 15251.

References

de Dear R, Schiller Brager G (1998) Developing an adaptive model of thermal comfort and preference, ASHRAE Trans., V.104(1a), pp145–167. Reprinted in Schiller Brager G (ed) (1998) Field studies of thermal comfort and adaptation. ASHRAE Tech Data Bull V.14(1):27–49

de Dear R, Brager G (2002) Thermal comfort in naturally ventilated buildings—revisions to ASHRAE Standard 55. Energy Build 34(6):549–561

EN 15251:2007–08 (2007) Criteria for the indoor environment. Beuth Verlag, Berlin

Humphreys M, Nicol F (2000) Outdoor temperature and indoor thermal comfort: raising the precision of the relationship for the 1998 ASHRAE database of field studies. ASHRAE Trans 206:485–492

ISSO Publicatie 74 (2004) Thermische Behaaglijkeid, ISSO, Rotterdam

Kalz DE, Hölzenbein F, Pfafferott J, Vogt G (2013) Nutzerzufriedenheit mit dem thermischen Komfort in Bürogebäuden mit Umweltenergiekonzepten. Bauphysik 35(6):377–391

Statistical Packages for the Social Sciences, Version 20.0 (SPSS) (2008) IBM Social Media Analytics, Armonk (NY)

van der Linden AC, Boersta AC, Raue AK, Kuvers SR, de Dear RJ (2006) Adaptive temperature
    limits: a new guideline in The Netherlands—a new approach for the assessment of building
    performance with respect to thermal indoor climate. Energy Build 38(1):8–17
Wagner A, Gossauer E, Moosmann C, Gropp Th, Leonhart R (2007) Thermal comfort and
    workplace occupant satisfaction—results of field studies in German low-energy office
    buildings. Energy Build 39(7):758–769

Chapter 5
Methodology for the Evaluation of Thermal Comfort in Office Buildings

Abstract The comfort standards define a methodology for the evaluation of thermal comfort for design purposes. Compared to this, a standardized evaluation of measurement campaigns should apply some constraints: The applied comfort approach need to be adapted for the cooling concept. Monitored room temperatures are evaluated on an hourly basis for both the upper and lower limits during occupancy only. The evaluation is carried out for 84 % of the building area. The deviation from the comfort temperature and its tolerance band is limited to 5 % and is determined for the entire summer period. A summer day is a day with a mean running temperature higher than 15 °C. Results should be illustrated in a comfort figure and as a thermal-comfort footprint.

Thermal comfort in office buildings is evaluated in accordance with two models defined in the European standard EN 15251:2007–08 (2007): the PMV and the adaptive-comfort model. In accordance with the defined comfort standard, a standardized evaluation for measurement campaigns is presented in order to evaluate thermal comfort in summer in office buildings under real operation.

In the context of a holistic design of a building and its energy concept, different parameters of the building physics and the HVAC systems are already considered and evaluated in detail during the planning phase. The design process considers the specific use of the building and technical equipment of offices and meeting rooms. Besides the climate conditions, interior heating and cooling loads are influenced notably by the behavior of users. As the loads vary in space and time, the designed HVAC system may not be able to fulfill the defined comfort requirements in all rooms during any time of occupancy. During the design of a building, users' behavior in terms of attendance and the operation of windows, sun shading and artificial lighting has to be anticipated carefully. In practice, nevertheless, the question often arises whether under real operation, the building will actually comply with the requirements defined in the planning phase with regard to thermal interior comfort in summer. Under certain circumstances, the perceived indoor environment may not meet user's expectations (See Chap. 4).

A metrological investigation offers the possibility to provide objective data and, therefore, to assess thermal comfort in buildings under operating conditions. Then

D. E. Kalz and J. Pfafferott, *Thermal Comfort and Energy-Efficient Cooling*
of Nonresidential Buildings, SpringerBriefs in Applied Sciences and Technology,
DOI: 10.1007/978-3-319-04582-5_5, © The Author(s) 2014

the effectiveness of the cooling and ventilation concept can be evaluated. Standard DIN EN 15251 (2007) presents indicators for an evaluation of interior thermal comfort by using office rooms representative for different zones in the building.

Furthermore, building occupants are a valuable source of information on building performance as well as indoor environmental quality and their effects on comfort and productivity. A large number of different studies have been conducted over recent decades, which focused on various aspects of the broad field of comfort, well-being, and health at workplace.

Some studies of individual buildings try to combine long-term monitoring of thermal comfort with post-occupancy evaluation and correlate these findings with the energy consumption for heating, cooling, and ventilation. Three important field studies on certain aspects of thermal comfort are:

- HOPE study: Questionnaire surveys were conducted with occupants of 96 apartment and 64 office buildings in Europe, providing information on how they felt and perceived their internal environment. Delivered energy use was assessed from records and energy bills, preferably collected over several years and averaged in order to assess an average annual consumption. However, no detailed breakdown of energy used for the HVAC system is available. Results just cover the building's total final and primary energy use. Perceived productivity was found to be better and absenteeism smaller in low-energy buildings. Too high temperatures in summer were reported to decrease perceived productivity. This correlation is less obvious in winter, but the productivity tends to decrease if the temperature is not perceived to be satisfactory. The occupants' perceived opportunity to control the environment correlates with the corresponding perceived comfort (Roulet 2001; Leaman and Bordass 2007).
- PROBE study: This study has undertaken post-occupancy surveys of 16 new commercial and public buildings, typically 2–3 years after completion. The purpose was to provide feedback on factors for success in the design, construction, operation, and use of buildings. The ventilation technologies in the buildings encompass air condition, night-ventilation, as well as natural and hybrid ventilation concepts. Energy data had been collected from monthly or quarterly invoices as well as manual and site meter readings without comprehensive monitoring, i.e., available records encompass the building's total fossil-fuel consumption and electrical energy use. However, findings of the occupant survey are not correlated to energy use or building operation (Cohen et al. 2001; Derbyshire 2001; Bordass et al. 2001a, b; and Leaman and Bordass 2001).
- EnOB study: Seventeen office buildings in Germany were assessed with a questionnaire in terms of thermal, visual, and acoustic comfort as well as indoor air quality and office layout (Gossauer 2008; Gossauer and Wagner 2008). Spot measurements of room temperature and relative humidity were conducted. Results of the post-occupancy evaluation were related to the energy concept of the buildings; however, values of energy use and efficiency were not gathered. The research shows that there are distinctive differences between the thermal-comfort voting in the summer and winter periods. Besides, it confirms that the

occupants' opportunity to influence their surrounding conditions and the perceived effectiveness of these interventions are crucial for perceived comfort and satisfaction.

5.1 Methodology for the Evaluation of Thermal Comfort

This chapter presents a detailed discussion of the various boundary conditions that need to be considered for the evaluation of interior thermal comfort in office buildings. The methodology is explained by using monitored nonresidential buildings in Europe as examples (building descriptions and analyses in Chap. 6).

Building category. Previous investigations revealed that the variety of heating and cooling concepts for the building stock and new constructions cannot be covered by just two categories in the current EN 15251:2007–08 standard, i.e., mechanically and non-mechanically cooled buildings (Chap. 4). Consequently, it is proposed to define five buildings standards, listed in Table 5.1:

(1) Buildings without cooling,
(2) Low-energy buildings with passive cooling,
(3) Low-energy buildings with air-based mechanical cooling,
(4) Low-energy buildings with water-based mechanical cooling,
(5) Buildings with mixed-mode cooling (combination of air-conditioning and air- or water-based mechanical cooling), and
(6) Buildings with air-conditioning.

Unfortunately, the national and international denotation of energy concepts for buildings is ambiguous. In the U.S., mixed-mode buildings are conceived as buildings that are mainly air-conditioned but use free ventilation of the office area during periods with favorable ambient conditions. The European perspective differs from that definition insofar as mixed-mode buildings employ cooling technologies with limited power (e.g., use of environmental heat sinks), but abstain from full air-conditioning.

Thermal-Comfort standards. The following categories are proposed for the evaluation of thermal comfort in summer:

- *Adaptive-comfort approach for low-energy buildings with passive or without cooling*: Again, the development of interior thermal comfort depends strongly on the behavior of the occupants and their use of the rooms, e.g., operation of windows, doors, and solar-shading system, the technical equipment of the rooms, the presence of occupants, use of the rooms as open plan or single office. Thermal comfort is evaluated in accordance with the adaptive approach of EN 15251:2007–08 (2007).
- *Adaptive-comfort approach for low-energy buildings with air-based mechanical cooling*: In these buildings, the level of adaptation and expectation is strongly related to outdoor climatic conditions. The application of an adaptive-comfort

Table 5.1 Categorization of building types for the evaluation of thermal comfort. See also Table 1.1 for system characteristics and requirements on the operation. Green: ventilation, blue: cooling, and gray: heat sink, compression chiller (CCH)

	No Cooling	Passive Cooling	Air-Based Cooling	Water-Based Cooling	Mixed-Mode Cooling	Air-Conditioning
Ventilation	Free ventilation over windows	Free ventilation over windows	Hybrid ventilation (over windows, exhaust or supply-/exhaust-ventilation system)	Hybrid ventilation (over windows, exhaust or supply-/exhaust-ventilation system) precooling of supply air	Free ventilation over windows air-conditioning	Hybrid ventilation (over windows and air-conditioning system)
Cooling	None	Free night-ventilation passive cooling technologies (e.g., solar shading, higher thermal mass of building, daylight concept)	Mechanical night-ventilation cooling of supply air by earth-to-air heat exchanger	Thermo-active building systems suspended cooling panels cooling of supply air	Cooling and dehumidification of supply air	Cooling and dehumidification of supply air

(continued)

Table 5.1 (continued)

	No Cooling	Passive Cooling	Air-Based Cooling	Water-Based Cooling	Mixed-Mode Cooling	Air-Conditioning
Heat sink	—	Ambient air	Ambient air	Surface near geothermal system, e.g., use of groundwater from well or surface near groundwater from borehole heat exchangers or use of air with cooling towers	Ambient air	Ambient air
Cooling generation	—	—	—	Direct cooling with cooling tower, groundwater well or borehole heat exchangers active cooling with compression chiller or reversible heat pump	Compression chiller	Compression chiller
Occupant	No adjustment of temperature set point for cooling, no dress code	No adjustment of temperature set point for cooling, no dress code	No adjustment of temperature set point for cooling, no dress code	Moderate influence of occupants to adjust temperature setpoints for cooling, usually no dress code	Strong influence of occupants on adjustment of temperature setpoint for cooling, usually dress code	Strong influence of occupants on adjustment of temperature setpoint for cooling, usually dress code
Comfort approach	*Adaptive approach* in accordance with EN 15251:2007–08 in dependence of the running mean ambient air temperature	*Adaptive approach* in accordance with EN 15251:2007–08 in dependence of the running mean ambient air temperature	*Adaptive approach* in accordance with EN 15251:2007–08 in dependence of the running mean ambient air temperature	*PMV approach* with constant setpoints independent from outdoor conditions in accordance with EN 15251:2007–08	*PMV approach* with constant setpoints independent from outdoor conditions in accordance with EN 15251:2007–08	*PMV approach* with constant setpoints independent from outdoor conditions in accordance with EN 15251:2007–08

model for the evaluation of thermal comfort in office buildings with free or mechanical night-ventilation is suitable. Occupants tolerate higher room temperatures at higher ambient air temperatures. The analysis of a field survey (see Chap. 4) and the resulting comfort temperature confirm the adaptive-comfort model of EN 15251:2007–08 (2007) for buildings with night-ventilation.

- *PMV-PPD comfort approach for low-energy buildings with water-based mechanical cooling*: Although users seem to adapt to the prevailing outdoor climate conditions, they expect a cooled interior environment and, therefore, have higher expectations on the interior thermal comfort. A field study revealed that users in these buildings tolerate only slightly higher room temperatures than the defined temperature setpoints in EN 15251 (2007) (see Chap. 4). Therefore, thermal comfort should be evaluated in accordance with the PMV model.
- *PMV-PPD comfort approach for buildings with air-conditioning*: Air-conditioned buildings provide a stable indoor environment. Therefore, user's expectations concerning indoor climate and, especially, room temperature are high. The user hardly influences the indoor climate.

Monitoring Campaigns. The measurement instrumentation used for evaluating thermal comfort and its location within the rooms should comply with the recommendations given in EN ISO 7726:2002–04 (2002). Measurements are to be made where occupants are known to spend most of their time and under representative weather conditions of warm seasons, advantageously at or above average outside temperatures during three warm months. The monitoring period for all measured parameters should be long enough to be representative. This depends on the time constant of the building and the prevailing weather conditions.

Nowadays, the building management system usually provides data for operative room temperature, relative air humidity, ambient air temperature, and plant-specific parameters (temperatures of supply system, operation time, ventilation rates, etc.). In new buildings, wall-mounted temperature sensors encapsulated in a ventilated enclosure are often available in all office rooms. Usually, these measurements can be taken as operative temperatures with adequate accuracy, as comparative measurements in the field have shown. Special care has only to be taken of large warm and cold surfaces. If no data are available, short-term monitoring with mobile measuring devices can be carried out for several weeks in summer. In addition to the indoor comfort parameters, the user behavior (operation of windows and solar protection) and physical characteristics of the space (surface temperatures, thermal efficiency of ventilation, velocity of supply air, etc.) can be recorded as well (Figs. 5.1, 5.2, 5.3).

Thermal-Comfort Assessment. Thermal-comfort assessments are determined separately for the summer and winter seasons in accordance with the comfort approaches of the European EN 15251:2007–08 standard. Evaluated are the numbers of hours during occupancy whenever the operative room temperatures exceed the defined upper and lower comfort limits of classes I, II, and III. Comfort ratings are analyzed in hours of exceedance during the time of occupancy.

Fig. 5.1 Mobile measurement devices for data acquisition, storage, and transmission of weather data (*left*) and microclimate at the façade. *Left* weather; *middle* and *right* microclimate at the façade (Fraunhofer ISE)

Fig. 5.2 Portable monitoring equipment for measuring thermal interior comfort: operative room temperature, air temperature, relative humidity, CO_2 concentration, and air velocity. *Right* logger for opening/closing of a window (Fraunhofer ISE)

Fig. 5.3 Installed monitoring equipment for operative room temperature and relative humidity (Fraunhofer ISE)

User Behavior. The allocation of the buildings to comfort classes is based entirely on long-term measurements. For that reason, user behavior (in terms of opening windows and using solar shading as well as their working activity and clothing level) is not being recorded.

Time of Occupancy. Thermal comfort is evaluated only during the time of occupancy, e.g., on weekdays from 8 a.m. to 7 p.m. Statutory holidays (e.g., Christmas, Easter) are considered, but not summer/winter vacation periods.

Temperature Drifts during Occupancy and Spatial Variations. For either comfort model, the operative temperature should be within the permissible ranges at all locations within the occupied zone of a space at all times. This means that the permissible range should cover both spatial and temporary variations, including fluctuations caused by the control system.

EN ISO 7730 (2005) and ASHRAE 55 (2004) restrict temperature drifts for different time periods, i.e., from 1.1 K per 15 min to 3.3 K per 4 h (on the basis of the PMV model). For these drifts, the thermal sensation can be estimated by using the PMV model. Even temperature drifts and ramps of $\pm$ 4 K per hour have no systematically significant effect on the objectively measured performance of subjects (Kolarik et al. 2007). However, experiments with fixed clothing insulation showed that continuous exposures to the increasing operative temperature for more than 4 h seemed to enhance the intensity of SBS symptoms. It is therefore recommended to avoid temperature drifts with rates of $\pm$ 4.4 K/h. Besides short-term (hourly) drifts, mid-(daily) and long-term (weekly) temperature drifts during the occupied hours of a space have to be taken into consideration. It is assumed that a day-to-day change in mean indoor temperatures of not more than 1 K with a cumulative change over a week of less than 3 K does not affect thermal comfort (Kolarik et al. 2008).

Spatial variations of the operative temperature can occur due to different surface temperatures (effective mean radiant temperature) and sources of heat and cold in a room (e.g., air inlets of air-conditioning systems). High quality insulation and low-e glazing—which are fundamental for energy-efficient office buildings—decrease the temperature differences between the façade(s) and the other surfaces of a space during winter and summer. This results in low-temperature heating and ventilation systems as well as in small necessary temperature differences for cooling. It can therefore be concluded that energy-efficient buildings provide—at first hand—low spatial temperature variations.

However, draft caused by big windows (even with U-values below 1 W/m^2 K) or air outlets as well as solar radiation have to be kept in mind. The latter can affect the local operative temperature in two ways: first through a high radiant temperature of the window or the shading system surface if radiation is absorbed to a significant extent; second, direct (single-sided) solar radiation on the human body has to be taken into account.

In naturally ventilated and passively cooled buildings, larger temporary variations of the operative temperature (temperature drifts) may, however, occur due to the fact that these buildings take advantage of their own thermal mass for heat accumulation. In the summer (cooling) season, heat (solar and internal loads) can be absorbed by the mass during the day, which is associated with a moderate temperature increase (up-drift). The thermal mass of the building is then cooled down at night either by ventilation (when outdoor temperatures are low) or by using the ground (water) as a direct heat sink. On the opposite, temperature down drifts can occur during winter time (which is not considered here). Allowing indoor temperatures to drift rather than maintaining them constant (which is common in most air-conditioned buildings) may be a feasible means of reducing the building's energy demand.

Figure 5.4 presents results on temperature drift during occupancy in European buildings, considering it during the morning and afternoon hours as well as throughout the entire day. Mostly, temperature drifts over the entire time of occupancy (8 a.m–6 p.m.) range between 0.5 and 3 K. This indicates that they are usually smaller than 1 K per hour. However, there are some exposed office rooms in the buildings where the temperature drifts exceed the limits due to occupant behavior and the use of the rooms. In conclusion, it can be stated that temperature drifts in European and German low-energy office buildings are mostly found to be around 1 K per hour and mainly between 1 and 3 K per day.

Acceptable Deviation in Location. Operative room temperatures and, finally, thermal-comfort ratings are evaluated separately for each room monitored in the building for the summer season. Obviously, the recorded temperatures vary significantly throughout the day within a building due to differing user behavior, room orientation, and presence of occupants (Kalz et al. 2009). EN 15251:2007–08 requires that the building meets the criteria of a specific (thermal-comfort) category if "the rooms representing 95 % of the building volume meet the criteria of the selected category". The authors do not believe that considering 95 % of the building for the evaluation of monitored data is a promising approach, since outliers dominate the evaluation procedure. These data do not characterize the entire building. There are exposed rooms in a building due to a wide range of reasons: occupants prefer higher operative room temperatures, no use of solar shading since the occupants prefer having a view outside, occupants are not present, rooms contain a lot of office equipment, a room has two external walls, rooms are occupied more densely than assumed during the design stage, etc. (Kalz et al. 2009).

An example: for a German office building, thermal comfort was evaluated for each single office room. Results are presented for the German building that employs mechanical night-ventilation in summer in order to cool down the building space (Fig. 5.5). All 15 monitored office rooms have the same size, the same use (office equipment) and the same geographical orientation. However, comfort results differ considerably between individual rooms, respecting thermal comfort requirements in accordance with class II for 100 % (room 1) or 90 % (room 13) of the time. Consequently, the differing development of the operative room temperature and the differing comfort ratings are attributed to occupant behavior only, i.e., presence of the occupants, opening of windows, use of solar shading, manual use of ventilation slats for night-ventilation. Therefore, the thermal-comfort condition of a building should be evaluated with reference to its

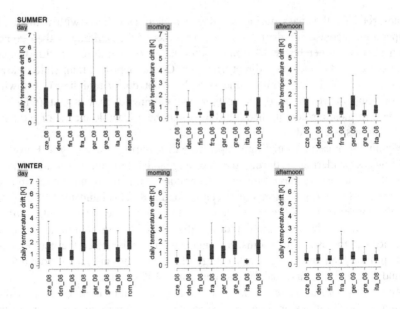

Fig. 5.4 Measured temperature drifts in summer and winter during occupancy (K). Considered are the daily drifts (*left*) as well as the ones during morning hours from 8 a.m. to 1 p.m. (*middle*) and during the afternoon hours from 1 to 6 p.m. (*right*). Results are given as boxplots, with 50 % of the values represented by the square, as well as minimum and maximum occurrences

typical thermal conditions, considering that there will be exposed rooms with temperatures above or below the average due to user behavior, user attendance, and room orientation.

Provided that the exceedance of the upper and lower comfort limits is a Gaussian variable, the standard deviation σ might be an appropriate scale unit. For design purposes, the recommendation is to use the established 95 % criterion as required by the current EN 15251:2007–08 standard. In order to preclude an overestimation of extremely high or low temperatures in monitoring campaigns, however, the recommendation is to use the (floor-area weighted) temperatures for 84 % of the building spaces according to the standard deviation σ. If there are less than five measurement points within the building, all rooms are considered for the comfort rating and not for the standard deviation.

Acceptable Deviation in Time. As recommended by EN 15251:2007–08, measured values of the operative room temperature are allowed to be outside the defined comfort boundaries during 5 % of working hours. This standard determines acceptable deviations on an annual, monthly, weekly, and even daily basis. However, findings suggest that a specification based on a monthly and weekly maximum of exceedance is not a promising approach, since it is too sensitive to a malfunction of the plant, improper operation, and user behavior. The exceedance of thermal comfort limits during moderate ambient conditions, e.g., periods during spring and autumn, is exclusively attributable to occupant behavior. The user has

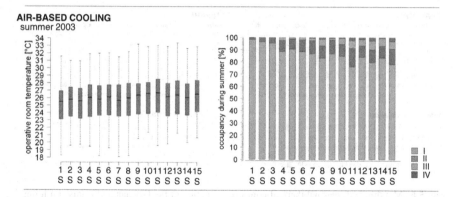

Fig. 5.5 Evaluation of room temperatures (°C) (*left*) and thermal comfort according to the adaptive-comfort approach of EN 15251:2007–08 (*right*) for an office building in Germany with air-based mechanical cooling (night-ventilation) during the relatively hot summer of 2003. Results are presented for 15 office rooms in one wing of the building (2 floors) with a southern orientation. Comfort classes I to IV in the right figure are from bottom to top

the opportunity to counteract the increasing operative room temperatures effectively by operating sun-shading devices or by opening windows. With respect to these results, it is proposed to determine thermal-comfort ratings on the basis of the entire summer season, and that the comfort class be allocated accordingly. For example, considering 1,358 working hours during summer, the number of tolerated working hours exceeding the comfort limits would be about 67 h per season.

Seasonal Exceedance of Comfort Boundaries. As postulated in EN 15251:2007–08, the defined comfort range has to be considered with regard to annual, monthly, weekly, and even daily exceedances. This includes a tolerance range of 5 %, which amounts to 13 h per month and 3 h per week, respectively. Within the framework of this study, comfort ratings were analyzed separately for the months of April to September and, additionally, for the 52 weeks of the year. An example for the monthly evaluation of thermal comfort is given in Fig. 5.6 for office buildings in Germany and Denmark.

The classification of the buildings to comfort classes differs considerably when the analysis is made on a monthly basis in comparison to a weekly or annual evaluation. Many buildings exceed the temperature limits for more than the tolerated number of hours per week—sometimes notably. However, this violation cannot be observed in every month. Interestingly, significant exceedances of the tolerance range mostly occur in the spring and autumn periods (April, May and September) when the running mean of the ambient air temperature varies between 13 and 18 °C, which therefore requires a setting of 23 °C in accordance with EN 15251:2007–08. It is questionable whether a week or month is representative for allocating the building to any thermal comfort class I to III. With respect to these results, it is proposed to determine thermal comfort ratings on the basis of the entire summer season and that the comfort class be allocated accordingly.

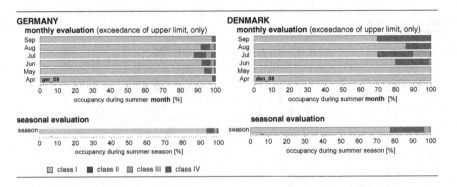

Fig. 5.6 Monthly and seasonal thermal-comfort evaluation of the German and Danish buildings, presented as thermal comfort footprints. *Note* only the exceedance of the upper comfort limit is considered. Squares for comfort classes I to IV from left to right

Summer and Winter Evaluation. The comfort standard EN 15251:2007–08 is not consistent with its definition of the summer and winter periods, i.e., the distinction of the upper comfort boundaries according to the seasons differs between the adaptive and the PMV-comfort approach. Fanger's thermal-comfort model (PMV-comfort approach) requires the input variables "metabolic rate" and "insulation level of clothing" (winter period 1.0 clo and summer period 0.5 clo). The prevailing ambient conditions are not considered in the model. Therefore, it is not explicit when Fanger's model refers to summer or winter conditions. The adaptive-comfort approach defines upper comfort boundaries for a running mean ambient air temperature of 10–30 °C and lower comfort boundaries for a temperature range of 15–30 °C.

Haldi and Robinson (2008) conclude from field studies that the clothing level can be reliably modeled by outdoor conditions, for example by using regressions on running mean ambient air temperature. As a result, clothing adaptation tends to be a more predictive strategy—the level being set at the beginning of the day, based on prior experience of outdoor thermal conditions. This relationship is expressed by a linear regression with good agreement (R^2 0.97) (Haldi and Robinson 2008). Therefore, a running mean ambient air temperature of 15 °C would result in a *clo* factor of 0.7 and at 22 °C in a clo factor of 0.5, which is the criterion for the summer period in accordance with ISO 7730:2005 (2005).

Though different studies on people's clothing behavior come to slightly different conclusions on the dependency between clo value and outdoor temperature, all studies show that the clo value converges to 1 for low (1–2 °C) and to 0.5 for high (22–27 °C) daily or running mean temperatures, respectively. These studies show that the clo value is 0.7 at approximately 15 °C outdoor temperature. Consequently, a reasonable "switching temperature" from winter to summer is an outdoor running mean temperature of 15 °C, since a clo value of 0.7 is typical office clothing with long-sleeved shirt but without jacket.

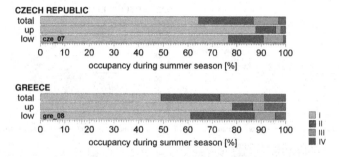

Fig. 5.7 Thermal comfort footprint for buildings in Greece and the Czech Republic: occupancy during summer season (%) if thermal comfort complies with classes I to IV; here for the adaptive-comfort approach of EN 15251. Exceedance of *upper comfort* limits only (*up*), of *lower comfort* limits only (*low*), *upper and lower comfort* limits are considered (*total*). Squares for comfort classes I to IV from left to right

The recommendation is to use a clear temperature reference for both the PMV and the adaptive-comfort approach. Heating mode and winter season are below an outdoor running mean temperature of 15 °C, while cooling mode and summer season are above 15 °C. The real cooling or heating time (energy [kWh]) may differ from the perceived summer or winter season (adaptation [clo]).

Building Classification. In accordance with the comfort criteria, the buildings are assigned to a comfort class I, II, or III, indicating the percentage of satisfied occupants. The requirement for a certain comfort class is fulfilled if at least 84 % of the recorded hourly temperature measurements remain within the defined comfort limit and its equivalent tolerance range. Comfort class II represents a "normal level of expectation and should be used for new buildings and renovations" (EN 15251).

Thermal Comfort Footprint. Comfort results for a building with its energy concept for heating, cooling, and ventilation are presented as *thermal comfort footprint* (Fig. 5.7), indicating the time of occupancy when thermal interior comfort complies with classes I to III. The period is given as a percentage of the total occupancy during summer. This fosters the comparison of the annual energy demand/consumption for heating and cooling with the simulated/monitored thermal comfort.

Presentation of Thermal-Comfort Results. As the "footprint" characterizes the building in a general matter, clients and building operators may not be able to understand the conclusion, especially the relevance of room temperatures exceeding the upper comfort limit in winter and the lower limit in summer. Therefore, we recommend to clearly state that the comfort diagram should be shown in addition to the footprint. Despite the building's categorization, the results of the thermal-comfort assessment should be presented for both the adaptive and the PMV-comfort approach. This will provide the client with data for the expected performance of the entire building concept Fig. (5.8)

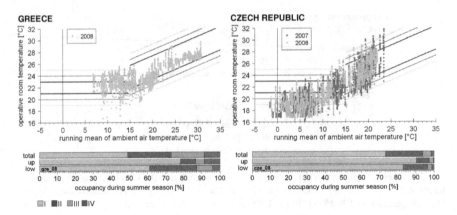

Fig. 5.8 Presentation of thermal comfort results for buildings in Greece and the Czech Republic: thermal comfort figure and thermal comfort footprint; here for the adaptive-comfort approach of EN 15251

5.2 Summary: Evaluation of Thermal Comfort in Office Buildings

Thermal room comfort is evaluated in accordance with two models of the European EN 15251:2007–08 standard: the PMV- and the adaptive-comfort model. In accordance with the defined comfort standard, a standardized evaluation of the measurement campaigns applies the following constraints (Kalz 2011):

- *Building category*: buildings with passive and air-based mechanical cooling are evaluated according to the adaptive-comfort approach of the EN 15251:2007–08 guideline. Buildings that condition the rooms and offices by means of water-based mechanical cooling and full air-conditioning are evaluated with respect to the PMV-comfort approach of EN 15251:2007–08. Evaluated are the numbers of hours during occupancy when the operative room temperatures exceed the defined upper and lower comfort limits of class II.
- *Occupancy*: Thermal comfort is evaluated only during the time of occupancy, e.g., on weekdays from 8 a.m. to 7 p.m. The periods of public holidays and vacations are not taken into account.
- *Building area*: Thermal comfort evaluation of a building under operation is carried out for at least 84 % (standard deviation) of the building area. However, during the design stage of the building and the technical plant, 95 % of the building area are required to meet the comfort class, based on the assumption of standardized occupant behavior.
- *Seasonal evaluation*: Thermal comfort, i.e., compliance with the defined comfort boundaries I to III, is determined for the entire summer period and not on a daily or weekly basis.

- *Definition of summer period*: Summer season should be defined as the period with a running mean ambient air temperature above 15 °C. Accordingly, the winter season is defined as the period with a value below 15 °C. The running mean ambient air temperature considers the history of the ambient conditions and therefore as well the development of the interior thermal conditions. Low-energy buildings usually employ considerable thermal storage capacities due to their exposed concrete ceilings.
- *Range of tolerance for comfort evaluation*: As recommended by EN 15251:2007–08, measured values of the operative room temperature during occupancy are allowed to be outside the defined comfort boundaries I to III during a maximum of 5 % of the working time in summer season. In other words, during 95–97 % of the occupancy time, the required thermal conditions are met.
- *Comfort class for thermal comfort*: Thermal comfort is evaluated in accordance with the defined comfort classes I to III (class I—high level of expectation, class II—normal level of expectation, class III—acceptable, moderate level of expectation, and class IV—values outside the criteria for the above categories).
- *Comfort class for humidity comfort*: Relative humidity is also evaluated for comfort classes I to III in accordance with EN 15251:2007–08. It should be noted that these limits are defined for the design of air-conditioning systems with dehumidification devices.
- *Presentation of thermal-comfort results*: Results of the monitoring campaigns are illustrated in a comfort figure and as a thermal-comfort "footprint". According to the required comfort model, hourly operative room temperature values (average) of the building during the time of occupancy are plotted against the running mean ambient air temperature.

References

ASHRAE Standard 55-2004 (2004) Thermal environmental conditions for human occupancy. American Society of Heating, Refrigerating and Air-Conditioning Engineers Inc., Atlanta

Bordass B, Cohen R, Standeven M and Leaman A (2001a) Assessing building performance in use 2: Technical performance of the Probe buildings. Build Res Inf 29(2):103–113

Bordass B, Cohen R, Standeven M, Leaman A (2001b) Assessing building performance in use 3: energy performance of the probe buildings. Build Res Inf 29(2):114–128

Cohen R, Standeven M, Bordass B, Leaman A (2001) Assessing building performance in use 1: the probe process. Build Res Inf 29(2):85–102

Derbyshire A (2001) Editorial probe in the UK context. Build Res Inf 29(2):79–84

DIN EN ISO 7726:2002–04 (2002) Ergonomics of the thermal environment—Instruments for measuring physical quantities. Beuth, Berlin

DIN EN ISO 7730:2005 (2005) Ergonomics of the thermal environment—Analytical determination and interpretation of thermal comfort using calculation of the PMV and PPD indices and local thermal comfort criteria. Beuth, Berlin

DIN EN 15251:2007–08 (2007) Criteria for the indoor environment. Beuth, Berlin

Gossauer E (2008) Nutzerzufriedenheit in Bürogebäuden—eine Feldstudie. Dissertation, University of Karlsruhe. Fraunhofer IRB Verlag, Stuttgart

Gossauer E, Wagner A (2008) Occupant satisfaction at workplaces—a Field study in office buildings. In: Proceedings of Windsor conference on air-conditioning and the low carbon cooling challenge, London Metropolitan University, Windsor, July 2008

Haldi F, Robinson D (2008) On the behavior and adaptation of office occupants. Build Environ 43(12):2163–2177

Kalz DE, Pfafferott J, Herkel S, Wagner A (2009) Building signatures correlating thermal comfort and low-energy cooling: in-use performance. Build Res Inf 37(4):413–432

Kalz DE (2011) Heating and Cooling Concepts Employing Environmental Energy and Thermo-Active Building Systems, System Analysis and Optimization. Dissertation, University of Karlsruhe. Fraunhofer IRB Verlag, Stuttgart

Kolarik J, Olesen BW, Toftum J and Mattarolo L (2007) Thermal Comfort, Perceived Air Quality and Intensity of SBS Symptoms During Exposure to Moderate Operative Temperature Ramps. In: proceedings of WellBeing Indoors-Clima conference, CD-ROM, Helsinki, Finland, 2007

Kolarik J, Toftum J, Olesen BW and Shitzer A (2008) Human subjects' perception of indoor environment and their office work performance during exposures to moderate operative temperature ramps. In: 11th International Conference on Indoor Air Quality and Climate, Copenhagen, Denmark

Leaman A, Bordass B (2001) Assessing building performance in use 4: the probe occupant surveys and their implications. Build Res Inf 29(2):129–143

Leaman A, Bordass B (2007) Are users more tolerant of 'green' buildings? Build Res Inf 35(6):662–673

Roulet CA (2001) Indoor environment quality in buildings and its impact on outdoor environment. Energy Build 33(3):183–191

Chapter 6
Thermal-Comfort Evaluation of Office Buildings in Europe

Abstract This chapter presents a comparative evaluation of thermal comfort according to the European standard EN 15251 standard in 8 European and another 34 German nonresidential buildings with different cooling concepts. Evidently, the comfort performance is not strongly affected by the type of environmental heat sink employed, provided that the heat sink is adequately dimensioned and well-operated. As expected, the room temperature in buildings with passive cooling or air-based lowenergy cooling is slightly higher at 25 °C and the occurring range of room temperatures is wider than in buildings with water-based low-energy cooling or air-conditioning at 23.5 °C. Monitoring results indicate that the buildings with radiant cooling and environmental-energy systems with lower cooling capacities are sensitive towards the applied control and operation algorithms as well as occupant behavior. An unexpected result is the wide range of comfort ratings within a given building. Obviously, both the building concept and the user behavior strongly affect the individual indoor environment. Convincing building build on passive cooling concepts in order to reduce cooling loads and to stabilize the room temperature and are characterized by the fact that users are enabled to influence their interior surroundings effectively.

This chapter presents a comparative evaluation of thermal comfort according to the European EN 15251:2007–08 standard in 8 European and another 34 German nonresidential buildings. The allocation of the buildings to comfort classes is based entirely on long-term monitoring campaigns. The European buildings are located in Finland, Denmark, France, Germany, the Czech Republic, Romania, Italia, and Greece, and therefore cover all main climate regions of Europe.

6.1 Description of the Investigated Buildings

Building Concept. In spite of different approaches for architecture and design, all of the office buildings in this study strive for a significantly reduced primary-energy use with carefully coordinated measures: high-quality building envelopes,

D. E. Kalz and J. Pfafferott, *Thermal Comfort and Energy-Efficient Cooling of Nonresidential Buildings*, SpringerBriefs in Applied Sciences and Technology, DOI: 10.1007/978-3-319-04582-5_6, © The Author(s) 2014

Table 6.1 Information on European and German buildings studied

Building + Building physics											Measurement	
Building	Location	Established	Retrofit (r)	Net floor area [m²]	U value ex. wall [W/(m² K)]	U value window [W/(m² K)]	g value window	A/V ratio [m⁻¹]	Glass to façade ratio [%]	Shading	Reference rooms	Measurement
No cooling (7)												
AA	GER	n/k	r	n/k	0.8	1.4	0.6	n/k	60	i	1	2W
AB	GER	n/k	r	11,915	0.26	1.4	0.58	n/k	32	n/k	1	2W
AC	GER	1996		1,000	0.47	1.8	0.6	0.55	70	e	5	4Y
AD	GER	1975	r	290	0.16	1.5	0.6	n/k	n/k	e	2	7W
AE	GER	1970		4,032	0.6	2.8	0.77	0.3	40	e	2	4W
AF	GER	n/k	r	n/k	0.5	1.2	0.78	n/k	50	i	6	2W
AG	GER	1960		4,243	0.53	2.9	0.7	0.29	40	n/k	2	2W
Passive cooling (5)												
AH	CZE	2005		230	0.2	1.2	0.7	0.4	20-80	i	1	1Y
AI	FRA	2003		3,900	0.43	2.2	n/k	0.1	40	i	9	1Y
AJ	GER	2001		13,150	0.23	1.4	0.58	0.31	25-55	e	5	2Y
AK	GER	n/k		n/k	0.45	1.2	0.34	n/k	54	b	1	1W
AL	GER	1953	r	30,570	0.61	3.0	0.7	0.23	70	i+e	14	3Y
Air-based cooling (10)												
AM	GER	1999		5,974	0.3	1.6	0.6	0.27	31	e	9	3Y
AN	GER	1950	r	986	0.14	0.8	0.52	0.49	20	i	1	3Y
AO	GER	2001		13,150	0.23	1.4	0.58	0.31	25	e	16	2Y
AP	GER	2002		8,762	0.35	1.4	0.4-0.7	0.25	31	e	20	6Y
AQ	GER	1999		1,000	0.2	1.1	0.6	0.40	44	e	2	3Y
AR	GER	2001		3,510	0.2	1.4	0.6	0.32	51	e	1	2Y
AS	GER	1968		2,544	0.46	1.4	0.64	0.32	n/k	e	2	4Y

(continued)

Table 6.1 (continued)

Building	Location	Established	Retrofit (r)	Net floor area [m²]	U value ex. wall [W/(m² K)]	U value window [W/ (m² K)]	g value window	A/V ratio [m⁻¹]	Glass to façade ratio [%]	Shading	Reference rooms	Measurement
AT	GER	2003		13,833	0.19	1.3	0.6	0.29	49	e	42	5Y
AU	GER	2002		8,120	0.2	1.4	0.5-0.6	0.36	n/k	e	7	2Y
AV	GER	2000		4,113	0.22	0.85	0.58	0.25	13-55	e	6	1Y
Water-based cooling (16)												
AW	GER	2002		2,151	0.17	0.80	0.5	0.37	37-49	i	9	3Y
AX	DEN	2002		21,199	0.20	1.12	0.58	0.3	31-60	i	3	1Y
AY	GER	2008		2,500	0.21	1.0	n/k	n/k	n/k	e	22	3Y
AZ	GER	2007		4,878	0.20	1-1.4	0.58	0.3	48	e	11	2Y
BA	GER	1976	r	7,640	0.46	1.6	n/k	0.36	30	e	3	2Y
BB	GER	1978	r	1,100	0.30	1.4	0.55	0.27	40-80	e	3	3Y
BC	GER	2002		6,911	0.13	0.84	0.5	0.22	48	e	20	2Y
BD	FIN	2005		6,900	0.22	1.1	0.38	n/k	n/k	i+e	12	1Y
BE	GER	2004		10,650	0.20	1.3	0.55	0.3	48-53	e	2	2Y
BF	GER	2008		4,527	0.21	1.5	0.5	0.29	30-95	e	22	3Y
BG	ITA	2007		752	0.34	1.4	0.6	0.6	20	e	3	1Y
BH	GER	2011		6,352	0.32	n/k	n/k	n/k	n/k	i	21	1Y
BI	GER	2009		19,500	n/k	n/k	n/k	n/k	n/k	e	100	1Y
BJ	GER	2009		4,500	0.33	1.1	0.5	n/k	n/k	e	14	1Y
BK	GER	2007		2,264	n/k	n/k	n/k	n/k	n/k	e	25	3Y
BL	GER	2001		1,347	0.11	0.8	0.6	0.34	35-85	e	22	2Y
Mixed-mode cooling (2)												
BM	GRE	1995		600	0.28	2.7	0.7	0.24	20-80	i+e	9	1Y
BN	ROU	1940		292	0.82	2.5	0.42	0.8	30-40	e	4	1Y
Air-conditioning (2)												
BO	GER	2000		6,880	0.2	1.3	0.6	n/k	64	e	87	2Y
BP	GER	86/92		10,984	n/k	n/k	n/k	0.39	n/k	e	9	2Y

(continued)

Table 6.1 (continued)

Building	Cooling delivery				Cooling generation					Ventilation			Heat recovery [%]
	Convector	TABS	CP	Air	Chiller	District cooling	Split unit	Ground	Ambient air	Free	Mechanical	Night-ventilation	
No cooling (7)													
AA									x	x			no
AB									x	x			no
AC									x	x			no
AD									x	x			no
AE									x	x			no
AF									x	x			no
AG									x	x			no
Passive cooling (5)													
AH				x					x	x		f	no
AI				x					x	x		f	55
AJ				x					x	x	x	f	no
AK				x					x	x		f	no
AL				x					x	x		f	no
Air-based cooling (10)													
AM				x					x	x	x	m	65
AN				x					x	x	x	m	80
AO				x					x	x	x	m	46
AP				x					x	x	x	m	no
AQ				x					x	x	x	m	80
AR				x					x	x	x	m	no
AS				x					x	x	x	m	56
AT				x					x	x	x	m	no

(continued)

Table 6.1 (continued)

Building	Cooling delivery				Cooling generation					Ventilation			Heat recovery [%]
	Convector	TABS	CP	Air	Chiller	District cooling	Split unit	Ground	Ambient air	Free	Mechanical	Night-ventilation	
AU		x		x					x	x	x	m	80
AV		x		x					x	x	x	m	68
Water-based cooling (16)													
AW	x							54		x	x		75
AX	x			x	955				CT		x	m	70
AY	x									x	x		73
AZ	x			x	155			88	CT	x	x		75
BA			x			x				x	x		no
BB			x					10		x	x		60
BC	x			x				140		x	x		65
BD	x			x		345				x	x		80
BE	x			x				250		x	x		70
BF	x			x	75					x	x		n/k
BG	x			x	64			x		x	x	f+m	52
BH	x			x	x				CT	x	x		n/k
BI	x	x			197					x	x		n/k
BJ	x			x	227			k	CT	x	x		n/k
BK	x									x	x		n/k
BL	x							7		x	x		73
Mixed-mode cooling (2)													
BM	x			x	40		x			x	x		k
BN				x			x			x	x	f	k

(continued)

Table 6.1 (continued)

Building	Cooling delivery			Cooling generation						Ventilation			Heat recovery [%]
	Convector	TABS	CP	Air	Chiller	District cooling	Split unit	Ground	Ambient air	Free	Mechanical	Night-ventilation	
Air-conditioning (2)													
BO				x	110						x		90
BP				x	400				CT	x	x		-

Area-to-volume ratio of the building (A/V), cooling tower (CT), water-based, ceiling-suspended cooling panel (CP), shading between the glass panes (b), exterior (e), interior (i), retrofit (r), weeks (W), years (Y). Numbers in the section "cooling generation" indicate the thermal cooling power [kW$_{therm}$] of the compression chiller and the environmental heat sinks

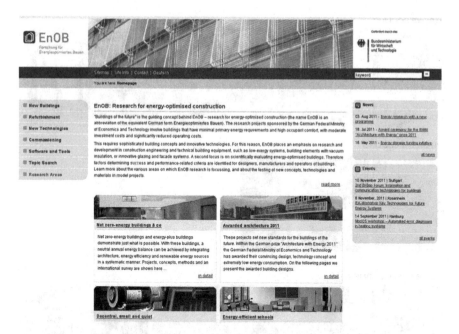

Fig. 6.1 Most of the German buildings studied belong to the EnOB research program (Research for energy-optimized construction (www.enob.info)). The research projects sponsored by the German Federal Ministry of Economics and Energy involve buildings with minimal primary-energy requirements and high occupant comfort, with moderate investment costs and significantly reduced operating costs

reduced solar heat gains (solar shading), sufficient thermal storage capacities, air-tight building envelopes in conjunction with hygienically necessary air-ventilation systems, and low-energy office equipment (reduced internal heat gains, daylight concepts). All buildings allow the user to influence the indoor environment with devices such as operable windows and sun-shading controls. Most of the office buildings consist of single or group offices, some also have open-plan offices. The buildings are described in Table 6.1 for each category of the cooling concepts, two demonstration buildings are described in detail (see building profiles in Sect. 6.2) (Fig. 6.1).

Energy Concept. The buildings studied are supposed to demonstrate the rational use of energy by means of innovative and soundly integrated technologies for the technical building services. The type of environmental and primary-energy use for heating, cooling, and ventilation is given in a schematic in the following building profiles. According to the main cooling system employed, the buildings are distinguished as proposed in Table 5.1 and Fig. 6.2.

Low-energy building with passive cooling (PC): A passive cooling concept covers all natural techniques of heat dissipation, overheating protection and related building-design techniques, providing thermal comfort without the use of mechanical

PASSIVE COOLING

AIR-BASED COOLING

WATER-BASED COOLING

AIR CONDITIONING

Fig. 6.2 Technologies for cooling of nonresidential buildings. *Passive cooling* free night-ventilation, daylighting concept solar shading (Fraunhofer ISE). *Air-based cooling* earth-to-air heat exchanger, mechanical night-ventilation (Fraunhofer ISE). *Water-based cooling* bore-hole heat exchangers, concrete core-conditioning (Fraunhofer ISE). *Air-conditioning* compression chiller, cooling distribution system, air-conditioning system (Fraunhofer ISE)

equipment and therefore auxiliary energy use. Passive cooling refers to preventing and modulating heat gains, including the use of natural heat sinks. For example, techniques are a well-designed building envelope and layout in high quality, solar control, internal gain control, and free night-ventilation. The design shall take the local microclimate and the building site into consideration. "Free night-ventilation" is simply a nonmechanical or passive means of providing ventilation through naturally occurring effects such as wind pressure on a building façade or stack effects within a building. During daytime, heat is stored in the structural elements of the building and is then rejected to the outdoor environment. However, only a certain amount of heat can be dissipated by night-ventilation due to the available nocturnal temperature level, the limited time, the practically feasible air-change rate, and the effectively usable heat-storage capacity of the building.

Low-energy building with air-based mechanical cooling (AMC): Besides the use of passive cooling techniques, night-ventilation is realized with a mechanical ventilation system. Is an exhaust-air ventilation system employed, indoor air is continuously exhausted to the outdoor environment. Fresh air is supplied through open windows or ventilation slats. Often the buildings employ a supply-and-exhaust air ventilation-system in order to make use of heat recovery in winter. Then indoor air is centrally exhausted to the outdoors and supply air is centrally sucked in and distributed to the individual rooms.

Low-energy building with water-based mechanical cooling (WMC): Hydronic radiant cooling systems encompass both integrated thermo-active building systems (TABS) and additive systems such as ceiling-suspended cooling panels. Due to a suitable construction method, TABS actively incorporate the structure (ceiling, wall, floor) and thermal storage into the energy management of the building. The broad range of TABS differs in dimension and spacing of the pipes, layer of thermal activation (surface-near or core), activated building component (ceiling, floor, wall), and implementation. The thermal properties of the constructions are predetermined by the vertical distance of the pipes to the surface, pipe spacing, floor and ceiling cover, pipe layout, volume flow and supply-water temperatures. Due to the large area for heat transfer, cooling is realized with relatively high supply-water temperatures between 16 and 22 °C. Under steady-state conditions, cooling capacities of 30–40 W/m^2 can be achieved. Near-surface systems can reach cooling rates of up to 70 W/m^2 under appropriate operating conditions. Higher cooling capacities are limited by the dew point of the room air temperature, as condensation would otherwise occur under the ceiling. The dew point is 15 °C at a room air temperature of 26 °C and a relative humidity of 50 %. TABS favor the use of environmental heat sinks in the close proximity of the building site, such as surface-near geothermal energy of the ground and groundwater, the use of rainwater and ambient air. Borehole heat exchangers, ground collectors, energy piles, earth-to-air heat exchangers and groundwater wells are technologies to harvest surface-near geothermal energy up to a depth of 120 m. Outdoor air can be used as heat sink by means of dry or wet cooling towers.

Building with mixed-mode cooling (MMC): Mixed-mode cooling refers to a hybrid approach to space conditioning that uses a combination of natural ventilation from operable windows (either manually or automatically controlled) or other passive inlet vents, and mechanical systems that provide air distribution and some form of cooling (Brager et al. 2007). Usually, it is a combination of natural or mechanical ventilation during permissible outdoor air conditions and full air-conditioning with dehumidification of the office space. Mixed-mode buildings may incorporate control strategies between mechanical and passive systems which may either be fully automated, manually controlled, or operated as some combination.

However, stringent classification and understanding of mixed-mode cooling is not consistent and agreed upon.

Building with full air-conditioning (AC): Air-conditioning of spaces refers to any form of cooling, heating, ventilation, or disinfection that modifies the condition of air. By considering the summer period, an air-conditioning system cools and dehumidifies the indoor air at almost constant setpoints throughout the time of occupancy, typically using a refrigeration cycle or sometimes evaporation.

Monitoring Campaigns. Various scientific teams carried out extensive long-term monitoring in fine time resolution of the building and plant performances for 1–5 years. The monitoring data consist of minute-by-minute and hourly measurements of temperature sensors and energy meters or manual heat-meter readings, if not stated otherwise. Thermal comfort is quantified by measurements of operative room temperatures and local meteorological conditions. In addition, useful cooling energy and electricity consumption were recorded hourly or by manual, weekly meter readings.

In general, data accumulation is associated with erroneous data due to the malfunctioning of sensors and outages. Raw data are processed before data evaluation, using a sophisticated method to remove erroneous values and outliers from the database. Data were recorded by building automation systems or by a stand-alone acquisition system. Thermal comfort is quantified by measurements of operative room temperatures and local meteorological conditions.

Usually, temperature sensors of class A or class B (PT100 or PT1000) were installed. The accuracy of temperature measurements is defined in DIN EN 60751:2009–05 (2009).

- Ambient condition: Although the monitoring equipment was designed and inspected carefully, there are errors in the ambient air temperature (e.g., insufficient protection against solar radiation). The order of measurement errors is estimated to be ±0.14–0.5 K.
- Room condition: Inaccuracies of the room temperature measurements are due to the sensor position, e.g., draught effect, height and position of the sensors or the impact of wall temperature on the measurement of the air temperature. The order of error is estimated to be ±0.13–0.5 K. The average measuring errors of temperature sensors is about ±0.2 K. High accuracy is necessary since the temperature variation of indoor conditions lies within a range of 1–5 K.

6.2 Monitoring Results for European Office Buildings

The building descriptions on the following double pages show:

- weather conditions throughout the year of monitoring: monthly mean ambient air temperature [°C] (markers), maximum and minimum daily temperature [°C], and monthly solar radiation (global horizontal) [kWh/m^2]. Data come from public weather stations or the building site.

- pictures of the buildings.
- a schematic of the respective energy concepts for heating, cooling, and ventilation.
- the thermal-comfort evaluation for the monitoring year: mean operative room temperature of the building [°C] during the time of occupancy, plotted against the running mean ambient air temperature [°C] in accordance with the comfort guideline EN 15251:2007–08. Gray and black lines indicate the upper and lower comfort boundaries I, II, and III. The thermal-comfort footprint for buildings indicates the time during occupancy in summer season [%] when thermal comfort complies with classes I to IV: class I (light green), class II (dark green), class III (orange), and outside the defined comfort classes (red). Exceedance of upper comfort limits only (up), of lower comfort limits only (low), upper and lower comfort limits are considered (total). Squares for comfort classes I to IV from left to right.
- the humidity comfort evaluation: evaluation of interior comfort in terms of relative humidity [%] during occupancy (8 a.m.–7 p.m., weekends are not considered) in accordance with standard DIN ISO 7730:2005. The thermal-comfort footprint for buildings indicates the time during occupancy in summer season [%] when thermal comfort complies with classes I to IV: class I (light green), class II (dark green), class III (orange), and outside the defined comfort classes (red). Squares for comfort classes I to IV from left to right. Results for summer season (S): light gray markers, results for winter season (W): dark gray markers.
- the building signature correlates the useful cooling energy [kWh_{therm}/m^2a], total primary-energy consumption of the building for heating, cooling, ventilation, and lighting [kWh_{prim}/m^2a], as well as the time of occupancy [%] when comfort class II is reached in terms of thermal and humidity comfort. The green rectangle (dotted line) represents the objective function for these parameters and the arrows indicate the direction of the optimum. The orange rectangle (solid line) represents results from the monitoring campaign. The scales are chosen individually for each criterion. The primary-energy factor for electricity is 2.3kWh_{prim}/kWh_{fin} for all projects presented. Objective functions for useful cooling energy are derived from simulation studies (see Chap. 7).

PASSIVE COOLING
La Rochelle France (46°15', -1°15', 5m)

CLIMATE

annual avg. temp. [°C]	12.8	
month. avg. max temp. [°C]	19.6	
month. avg. min temp. [°C]	4.4	
heating degree days [Kday]	2,278	
cooling degree days [Kday]	n/k	
design indoor temp. H [°C]	19	
design indoor temp. C [°C]	26	

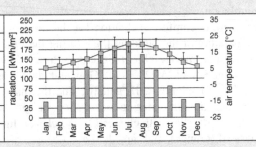

INFORMATION ON BUILDING AND USE

occupancy	university
number of occupants	15+1500
utilization	8am-6pm
completion	2003
refurbishment	-
number of floors	3
total floor area [m²]	3,900
total conditioned area [m²]	3,900
total volume [m³]	18,600
area -to-volume ratio [m⁻¹]	0.1

Source: University of La Rochelle

BUILDING ENVELOPE

shading system	interior blinds, manual operation, shading factor n/k
U-values [W/(m²K)]	exterior wall: 0.43 I window: 2.2 I roof: 0.36 I avg. value of building: 0.75
window	2-pane low -e glazing I g -value: 0.44 I area: 880m² I window-façade-ratio: 42%

COOLING CONCEPT

environmental heat sink	AA
energy carrier	-
cooling system	NV-f
power of system [kW therm]	-
distribution system	-

VENTILATION CONCEPT

operable windows	yes
night -ventilation	f
mechanical ventilation	yes
air -change rate [h⁻¹]	1.0
dehumidification of air	no
pre -cooling of air	no

ambient air (AA) I free (f) I mechanical ventilation (MV)
night -ventilation (NV)

THERMAL COMFORT PERFORMANCE IN SUMMER

year(s) of monitoring	2008-09
year of evaluation	2008-09
number of rooms	9
interval of measurements	60min
ambient air temperature	at building
design temperature [°C]	26
adaptive, class II (up)[a]	96%
adaptive, class II (low)[b]	38%
avg. room temperature [°C][c]	22–25.5
temp. drift day [K][c]	0.5–1.5
POE[d]	yes
humidity comfort, class II	44%

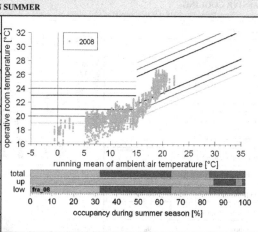

MONITORING RESULTS AND BUILDING SIGNATURE (period 7/08 to 09/09)

COOLING	
useful energy [kWh therm /m²a]	0
final energy [kWh fin/m²a]	0
primary energy [kWh prim/m²a]	0
HEATING	
useful energy [kWh therm /m²a]	74.3
final energy [kWh fin/m²a]	85.3
primary energy [kWh prim /m²a]	97.6
VENTILATION	
final energy [kWh fin/m²a]	n/k
primary energy [kWh prim /m²a]	n/k
LIGHTING	
final energy [kWh fin/m²a]	n/k
primary energy [kWh prim /m²a]	n/k
APPLIANCES/PLUG LOADS	
final energy [kWh fin/m²a]	n/k
TOTAL BUILDING	
final energy [kWh fin/m²a]	n/k
primary energy [kWh prim /m²a]	n/k
onsite generation of energy	no

(a) upper comfort boundaries, (b) lower
comfort boundaries, (c) during occupancy, (d)
post-occupancy evaluation

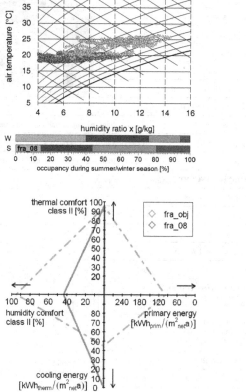

PASSIVE COOLING
Prague, Czech Republic (50°0', 14°34', 269m)

CLIMATE

annual avg. temp. [°C]	10	
month. avg. max temp. [°C]	18.9	
month. avg. min temp. [°C]	1.6	
heating degree days [Kday]	3,677	
cooling degree days [Kday]	19	
design indoor temp. H [°C]	20	
design indoor temp. C [°C]	26	

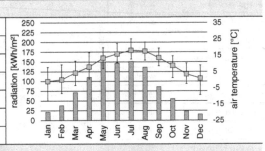

INFORMATION ON BUILDING AND USE

occupancy	tenement
number of occupants	4
utilization	9am-6pm
completion	2005
refurbishment	-
number of floors	2
total floor area [m²]	230
total conditioned area [m²]	230
total volume [m³]	562
area-to-volume ratio [m⁻¹]	0.4

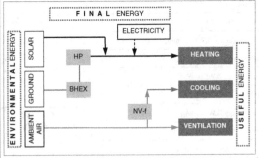

Source: Czech Technical University, Prague

BUILDING ENVELOPE

shading system	interior blinds, manual operation, shading factor n/k
U-values [W/(m²K)]	exterior wall: 0.2 I window: 1.4 I roof: 0.15 I avg. value of building: n/k
window	2-pane glazing I g-value: 0.4 I area: 54m² I window-façade-ratio: 20-80%

COOLING CONCEPT

environmental heat sink	AA
energy carrier	E
cooling system	NV-f, fans
power of system [kW_therm]	-
distribution system	air

VENTILATION CONCEPT

operable windows	yes
night-ventilation	f
mechanical ventilation	no
air-change rate [h⁻¹]	0.5
dehumidification of air	no
pre-cooling of air	no

ambient air (AA) I borehole heat exchanger (BHEX) I electricity (E) I free (f) I heat pump (HP) I night-ventilation (NV)

THERMAL COMFORT PERFORMANCE IN SUMMER

year(s) of monitoring	2008
year of evaluation	2008
number of rooms	1
interval of measurements	1 min
ambient air temperature	at building
design temperature [°C]	26
adaptive, class II (up)[a]	97%
adaptive, class II (low)[b]	96%
avg. room temperature [°C][c]	23-26
temp. drift day [K][c]	1.0-2.8
POE[d]	no
humidity comfort, class II[e]	n/k

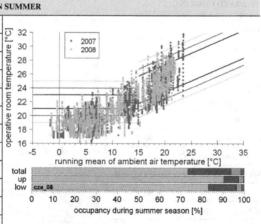

MONITORING RESULTS AND BUILDING SIGNATURE (period 7/08 to 04/09)

COOLING

useful energy [kWh$_{therm}$/m²a]	0
final energy [kWh$_{fin}$/m²a]	0
primary energy [kWh$_{prim}$/m²a]	0

HEATING

useful energy [kWh$_{therm}$/m²a]	56.0
final energy [kWh$_{fin}$/m²a]	68.0
primary energy [kWh$_{prim}$/m²a]	170.0

VENTILATION

final energy [kWh$_{fin}$/m²a]	0
primary energy [kWh$_{prim}$/m²a]	0

LIGHTING

final energy [kWh$_{fin}$/m²a]	4.6
primary energy kWh$_{prim}$/m²a]	11.5.

APPLIANCES/PLUG LOADS

final energy [kWh$_{fin}$/m²a]	n/k

TOTAL BUILDING

final energy [kWh$_{fin}$/m²a]	72.6
primary energy [kWh$_{prim}$/m²a]	181.5
onsite generation of energy	no

(a) upper comfort boundaries, (b) lower comfort boundaries, (c) during occupancy, (d) post-occupancy evaluation, (e) not measured

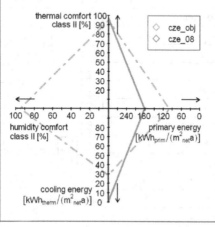

AIR-BASED COOLING
Freiburg Germany (47°99',7°84', 278m)

CLIMATE

annual avg. temp. [°C]	12.2	
month. avg. max temp. [°C]	25.0	
month. avg. min temp. [°C]	1.6	
heating degree days [Kday]	3,165	
cooling degree days [Kday]	-	
design indoor temp. H [°C]	20	
design indoor temp. C [°C]	26	

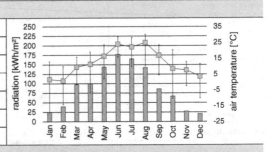

INFORMATION ON BUILDING AND USE

occupancy	office
number of occupants	400
utilization	8am-6pm
completion	2001
refurbishment	-
number of floors	3
total floor area [m²]	13,150
total conditioned area [m²]	6,474
total volume [m³]	64,322
area-to-volume ratio [m⁻¹]	0.31

Source: Fraunhofer ISE.

BUILDING ENVELOPE

shading system	exterior venetian blinds, automatic operation, shading factor 0.2
U-values [W/(m²K)]	exterior wall: 0.23 I window: 1. 4 I roof: 0. 34 I avg. value of building: 0.43
window	2-pane glazing I g-value: 0.58I area:—I window-façade-ratio: 28-50%

COOLING CONCEPT

environmental heat sink	AA
energy carrier	E
cooling system	-
power of system [kW$_{therm}$]	-
distribution system	air

VENTILATION CONCEPT

operable windows	yes
night-ventilation	yes
mechanical ventilation	yes
air-change rate [h⁻¹]	1
dehumidification of air	no
pre-cooling of air	no

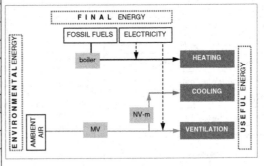

ambient air (AA) I electricity (E) I mechanical (m) I mechanical ventilation (MV) I night-ventilation (NV)

THERMAL COMFORT PERFORMANCE IN SUMMER

year(s) of monitoring	2002-2003
year of evaluation	2003
number of rooms	16
interval of measurements	5min
ambient air temperature	at building
design temperature [°C]	26
adaptive, class II (up)[a]	91%
adaptive, class II (low)[b]	95%
avg. room temperature [°C][c]	23.5-28
temp. drift day [K][c]	1.5-2.5
POE[d]	yes
humidity comfort, class II[e]	n/k

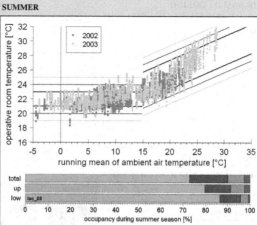

MONITORING RESULTS AND BUILDING SIGNATURE (period 1/03 to 12/03)

COOLING

useful energy [kWh$_{therm}$ /m²a]	18.4
final energy [kWh$_{fin}$ /m²a]	4.1
primary energy [kWh$_{prim}$/m²a]	10.6

HEATING

useful energy [kWh$_{therm}$ /m²a]	57.6
final energy [kWh$_{fin}$ /m²a]	6.2
primary energy [kWh$_{prim}$/m²a]	47.2

VENTILATION

final energy [kWh$_{fin}$ /m²a]	8.0
primary energy [kWh$_{prim}$/m²a]	20.8

LIGHTING

final energy [kWh$_{fin}$ /m²a]	2.8
primary energy [kWh$_{prim}$/m²a]	7.3

APPLIANCES/PLUG LOADS

final energy [kWh$_{fin}$/m²a]	n/k

TOTAL BUILDING

final energy [kWh$_{fin}$ /m²a]	76.9
primary energy [kWh$_{prim}$/m²a]	85.9
onsite generation of energy	yes

(a) upper comfort boundaries, (b) lower
comfort boundaries, (c) during occupancy, (d)
post-occupancy evaluation, (e) not measured

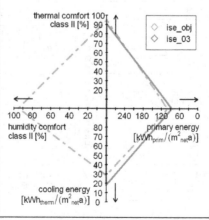

AIR-BASED COOLING
Freiburg Germany (47°99', 7°84', 278m)

CLIMATE

annual avg. temp. [°C]	11.7	
month. avg. max temp. [°C]	22.0	
month. avg. min temp. [°C]	-0.8	
heating degree days [Kday]	2,698	
cooling degree days [Kday]	n/k	
design indoor temp. H [°C]	20	
design indoor temp. C [°C]	26	

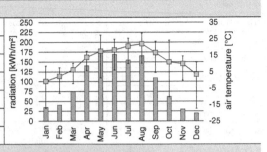

INFORMATION ON BUILDING AND USE

occupancy	office
number of occupants	400
utilization	6am-8pm
completion	2003
refurbishment	-
number of floors	6
total floor area [m²]	13,833
total conditioned area [m²]	13,833
total volume [m³]	53,629
area-to-volume ratio [m⁻¹]	0.29

Source: Solar Info Center, Freiburg

BUILDING ENVELOPE

shading system	exterior venetian blinds, automatic operation, shading factor 0.2
U-values [W/(m²K)]	exterior wall: 0.19 I window: 1.3 I roof: 0.19 I avg. value of building:
window	2-pane glazing I g-value: 0.60 I area: — I window-façade-ratio: 33-49%

COOLING CONCEPT

environmental heat sink	AA
energy carrier	E
cooling system	NV-m
power of system [kW_therm]	-
distribution system	air

VENTILATION CONCEPT

operable windows	yes
night-ventilation	m
mechanical ventilation	yes
air-change rate [h⁻¹]	1
dehumidification of air	no
pre-cooling of air	no

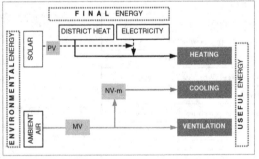

ambient air (AA) I electricity (E) I mechanical (m) I mechanical ventilation (MV) I nigh-ventilation (NV) I photovoltaic (PV)

THERMAL COMFORT PERFORMANCE IN SUMMER

year(s) of monitoring	2005-2010	
year of evaluation	2009	
number of rooms	10	
interval of measurements	5min	
ambient air temperature	at building	
design temperature [°C]	26	
adaptive, class II (up)[a]	93	
adaptive, class II (low)[b]	97	
avg. room temperature [°C][c]	25-27.5	
temp. drift day [K][c]	2.0-4.0	
POE[d]	yes	
humidity comfort, class II	98	

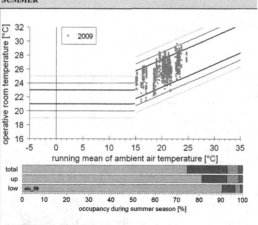

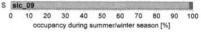

MONITORING RESULTS AND BUILDING SIGNATURE (period 1/09 to 12/09)

COOLING	
useful energy [kWh therm /m²a]	0.0
final energy [kWh fin /m²a]	0.0
primary energy kWh prim /m²a]	0.0
HEATING	
useful energy [kWh therm /m²a]	28.8
final energy [kWh fin /m²a]	32.0
primary energy [kWh prim /m²a]	26.2
VENTILATION	
final energy [kWh fin /m²a]	1.7
primary energy [kWh prim /m²a]	4.4
LIGHTING	
final energy [kWh fin /m²a]	11.0
primary energy [kWh prim /m²a]	28.6
APPLIANCES/PLUG LOADS	
final energy [kWh fin /m²a]	n/k
TOTAL BUILDING	
final energy [kWh fin /m²a]	47.7
primary energy [kWh prim /m²a]	67.1
onsite generation of energy	yes

(a) upper comfort boundaries, (b) lower
comfort boundaries, (c) during occupancy, (d)
post-occupancy evaluation

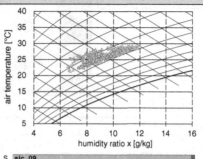

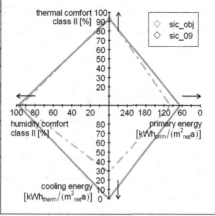

WATER-BASED COOLING
Karlsruhe Germany (49°0', 8°24', 115m)

CLIMATE

annual avg. temp. [°C]	10.3	
month. avg. max temp. [°C]	19.6	
month. avg. min temp. [°C]	1.2	
heating degree days [Kday]	3,264	
cooling degree days [Kday]	n/k	
design indoor temp. H [°C]	20	
design indoor temp. C [°C]	26	

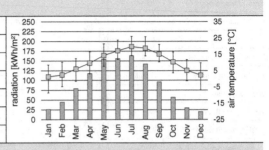

INFORMATION ON BUILDING AND USE

occupancy	office
number of occupants	50
utilization	7am-11pm
completion	1978
refurbishment	2005
number of floors	2
total floor area [m²]	1,390
total conditioned area [m²]	1,111
total volume [m³]	4,910
area-to-volume ratio [m⁻¹]	0.27

Source: Patrick Beuchert, Karlsruhe

BUILDING ENVELOPE

shading system	exterior venetian blinds, automatic operation, shading factor 0.2
U-values [W/(m²K)]	exterior wall: 0.3 I window: 1.4 I roof: 0.19 I avg. value of building: 0.54
window	solar control glazing I g-value: 0.55 I area: 473m² I window-façade-ratio: 20-87%

COOLING CONCEPT

environmental heat sink	AA, GR
energy carrier	E
cooling system	NV-f, BHEX
power of system [kWtherm]	10
distribution system	air, CP-w

VENTILATION CONCEPT

operable windows	yes
night-ventilation	f
mechanical ventilation	yes
air-change rate [h⁻¹]	1
dehumidification of air	no
pre-cooling of air	yes

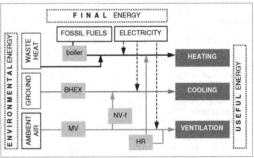

ambient air (AA) I borehole heat exchangers (BHEX) I electricity (E)
free (f) I ground (GR) I heat recovery (HR) I mechanical ventilation (MV)
I nigh-ventilation (NV) I water-driven, ceiling suspended cooling panels
(CP-w)

THERMAL COMFORT PERFORMANCE IN SUMMER

year(s) of monitoring	2008-10
year of evaluation	2008
number of rooms [a]	10
interval of measurements	5-min
ambient air temperature	public WS[b]
design temperature [°C]	26
adaptive, class II (up)[c]	88%
adaptive, class II (low)[d]	97%
avg. room temperature [°C][e]	23-27
temp. drift day [K][e]	1.5-4.0
POE[f]	yes
humidity comfort, class II	94%

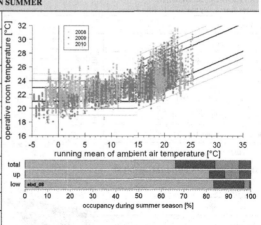

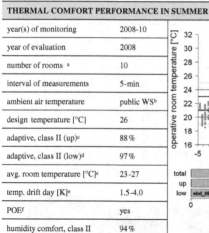

MONITORING RESULTS AND BUILDING SIGNATURE (period 1/08 to 12/08)

COOLING	
useful energy [kWh therm /m²a]	20.3
final energy [kWh fin /m²a]	5.9
primary energy kWh prim /m²a]	14.6

HEATING	
useful energy [kWh therm /m²a]	79.9
final energy [kWh fin /m²a]	98.3
primary energy [kWh prim /m²a]	102.0

VENTILATION	
final energy [kWh fin /m²a]	12.2
primary energy [kWh prim /m²a]	30.5

LIGHTING	
final energy [kWh fin /m²a]	23.5
primary energy [kWh prim /m²a]	59.1

APPLIANCES/PLUG LOADS	
final energy [kWh fin /m²a]	n/k

TOTAL BUILDING	
final energy [kWh fin /m²a]	140.0
primary energy [kWh prim /m²a]	206.3
onsite generation of energy	no

(a) open-plan office and normal offices, (b) public weather station in city, (c) upper comfort boundaries, (d) lower comfort boundaries, (e) during occupancy, (f) post-occupancy evaluation

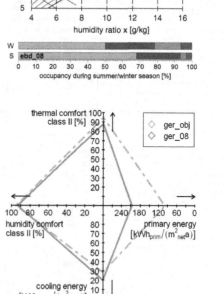

WATER-BASEDCOOLING
Porretta Terme Italy (44°16' , 10°97' , 349m)

CLIMATE

annual avg. temp. [°C]	12.3
month. avg. max temp. [°C]	22.6
month. avg. min temp. [°C]	1.2
heating degree days [Kday]	2,648
cooling degree days [Kday]	122
design indoor temp. H [°C]	20
design indoor temp. C [°C]	25

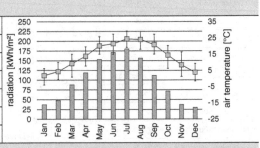

INFORMATION ON BUILDING AND USE

occupancy	civic center + office
number of occupants	20
utilization	8am - 12pm
completion	1970
refurbishment	2007
number of floors	2
total floor area [m²]	752
total conditioned area [m²]	580
total volume [m³]	1,930
area -to-volume ratio [m⁻¹]	0.6

Source: Politecnico di Milano

BUILDING ENVELOPE

shading system	exterior roller blinds, manual operation, shading factor 0.15
U-values [W/(m²K)]	exterior wall: 0.34 I window: 1.4 I roof: 0.35 I avg. value of building: 0.44
window	2-pane low-e glazing I g-value: 0.6 I area: 54m² I window-façade-ratio: 9-16%

COOLING CONCEPT

environmental heat sink	AA, GR
energy carrier	E
cooling system	NV, HP
power of system [kW$_{therm}$]	64
distribution system	coils, CV

VENTILATION CONCEPT

operable windows	yes
night-ventilation	f+m
mechanical ventilation	yes
air-change rate [h⁻¹]	1
dehumidification of air	no
pre -cooling of air	no

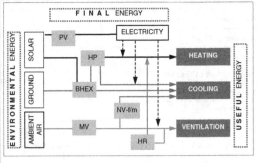

ambient air (AA) I borehole heat exchanger (BHEX) I convector (CV)electricity (E) I free (f) I ground (GR) I heat pump (HP) I heat recovery (HR) I mechanical (m) I mechanical ventilation (MV) I night-ventilation (NV) I photovoltaic (PV)

THERMAL COMFORT PERFORMANCE IN SUMMER

year(s) of monitoring	2008 -09
year of evaluation	2008 -09
number of rooms	3
interval of measurements	15min
ambient air temperature	public WS[a]
design temperature [°C]	25
adaptive, class II (up)[b]	100 %
adaptive, class II (low)[c]	88 %
avg. room temperature [°C][d]	24.5 -27.0
temp. drift day [K][d]	0.5 -1.5
POE[e]	yes
humidity comfort, class II	68 %

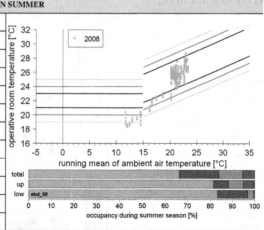
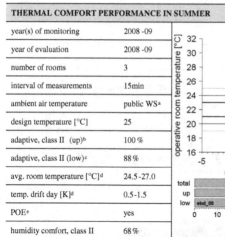

MONITORING RESULTS AND BUILDING SIGNATURE (period 7/08 to 9/08)

COOLING	
useful energy [kWh_therm /m²a]	30.5
final energy [kWh_fin /m²a]	25.0
primary energy [kWh_prim /m²a]	62.5
HEATING	
useful energy [kWh_therm /m²a]	88.2
final energy [kWh_fin /m²a]	64.1
primary energy [kWh_prim /m²a]	160.3
VENTILATION	
final energy [kWh_fin /m²a]	5.5
primary energy [kWh_prim /m²a]	13.8
LIGHTING	
final energy [kWh_fin /m²a]	1.8
primary energy [kWh_prim /m²a]	4.5
APPLIANCES/PLUG LOADS	
final energy [kWh_fin /m²a]	0.9
TOTAL BUILDING	
final energy [kWh_fin /m²a]	97.1
primary energy [kWh_prim /m²a]	242.8
onsite generation of energy	no

(a) public weather station, (b) upper comfort
boundaries, (c) lower comfort boundaries, (d)
during occupanc, (e) post-occupancy y
evaluation

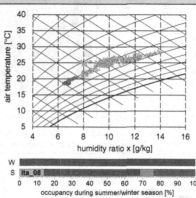

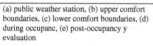

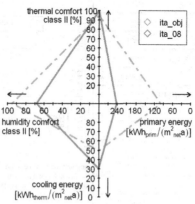

WATER-BASED COOLING
Copenhagen Denmark (55°67', 12°67', 24m)

CLIMATE

annual avg. temp. [°C]	7.7	
month. avg. max temp. [°C]	16.4	
month. avg. min temp. [°C]	-1	
heating degree days [Kday]	2,906	
cooling degree days [Kday]	n/k	
design indoor temp. H [°C]	20	
design indoor temp. C [°C]	25	

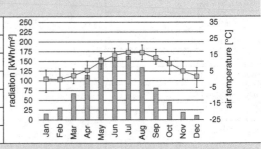

INFORMATION ON BUILDING AND USE

occupancy	office
number of occupants	1.700
utilization	8am-6pm
completion	2002
refurbishment	-
number of floors	5
total floor area [m²]	21,199
total conditioned area [m²]	18,726
total volume [m³]	71,533
area -to-volume ratio [m⁻¹]	0.3

Source: Danish Technical University, Lyngby

BUILDING ENVELOPE

shading system	interior venetian blinds, manual operation, shading factor n/k
U-values [W/(m²K)]	exterior wall: 0.2 I window: 1.12 I roof: 0.2 I avg. value of building: n/k
window	2-pane glazing I g-value: n/kI area: 302m² I window-façade-ratio: 30-60%

COOLING CONCEPT

environmental heat sink	AA
energy carrier	E
cooling system	chiller
power of system [kW_therm]	955
distribution system	coils + CP-w

VENTILATION CONCEPT

operable windows	no
night-ventilation	no
mechanical ventilation	yes
air -change rate [h⁻¹]	1.4
dehumidification of air	no
pre-cooling of air	yes

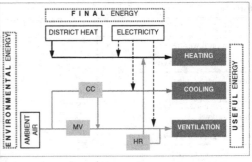

ambient air (AA) I compression chiller (CC) I water-driven, ceiling suspended cooling panels (CP-w I electricity (E) I heat recovery (HR) mechanical ventilation (MV)

THERMAL COMFORT PERFORMANCE IN SUMMER

year(s) of monitoring	2008
year of evaluation	2008
number of rooms	3
interval of measurements	15min
ambient air temperature	at building
design temperature [°C]	25
PMV, class II (up)[a]	92%
PMV, class II (low)[b]	89%
avg. room temperature [°C][c]	23-25.5
temp. drift day [K][c]	0.8-1.5
POE[d]	yes
humidity comfort, class II	97%

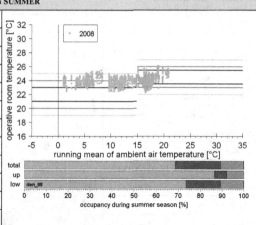

MONITORING RESULTS AND BUILDING SIGNATURE (period 1/08 to 12/08)

COOLING	
useful energy [kWh therm /m²a]	n/k
final energy [kWh fin /m²a]	n/k
primary energy [kWh prim /m²a]	8

HEATING	
useful energy [kWh therm /m²a]	n/k
final energy [kWh fin /m²a]	n/k
primary energy [kWh prim /m²a]	76.8

VENTILATION	
final energy [kWh fin /m²a]	n/k
primary energy [kWh prim /m²a]	5.5

LIGHTING	
final energy [kWh fin /m²a]	n/k
primary energy [kWh prim /m²a]	n/k

APPLIANCES/PLUG LOADS	
final energy [kWh fin /m²a]	n/k

TOTAL BUILDING	
final energy [kWh fin /m²a]	n/k
primary energy [kWh prim /m²a]	150.6
onsite generation of energy	no

(a) upper comfort boundaries, (b) lower
comfort boundaries, (c) during occupancy, (d)
post-occupancy evaluation

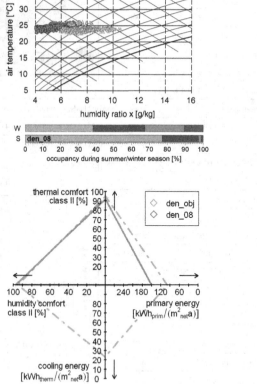

WATER-BASED COOLING
Turku Finland (55°67', 12°67', 24m)

CLIMATE

annual avg. temp. [°C]	5	
month. avg. max temp. [°C]	16.5	
month. avg. min temp. [°C]	-6.6	
heating degree days [Kday]	4,115	
cooling degree days [Kday]	194	
design indoor temp. H [°C]	21	
design indoor temp. C [°C]	25	

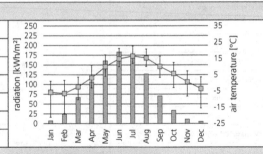

INFORMATION ON BUILDING AND USE

occupancy	office
number of occupants	135
utilization	8am-5pm
completion	2005
refurbishment	-
number of floors	5
total floor area [m²]	6,900
total conditioned area [m²]	6,900
total volume [m³]	34,000
area-to-volume ratio [m⁻¹]	n/k

Source: YIT, Finland

BUILDING ENVELOPE

shading system	exterior venetian blinds, automatic operation, shading factor n/k
U-values [W/(m²K)]	exterior wall: 0.22 I window: 1.1 I roof: 0.14 I avg. value of building: n/k
window	n/k I g-value: 0.38 I area: n/k I window-façade-ratio: n/k

COOLING CONCEPT

environmental heat sink	-
energy carrier	-
cooling system	district cool
power of system [kW_therm]	345
distribution system	air, CP-w

VENTILATION CONCEPT

operable windows	yes
night-ventilation	no
mechanical ventilation	yes
air-change rate [h⁻¹]	2.2
dehumidification of air	no
pre-cooling of air	yes

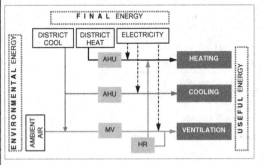

air handling unit (AHU) I water-driven, ceiling suspended cooling panels (CP-w) I heat recovery (HR) I mechanical ventilation (MV)

THERMAL COMFORT PERFORMANCE IN SUMMER

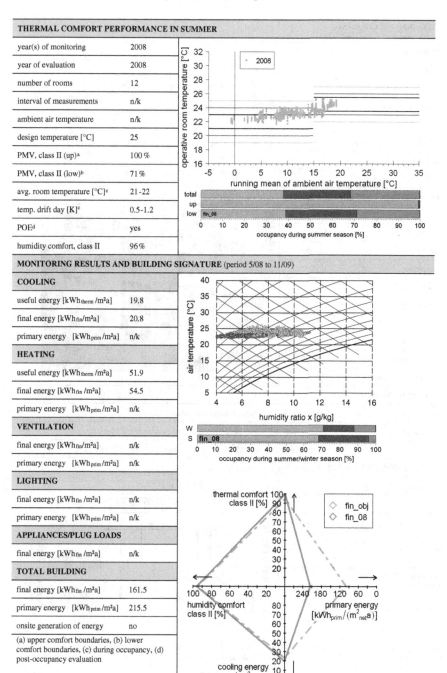

year(s) of monitoring	2008
year of evaluation	2008
number of rooms	12
interval of measurements	n/k
ambient air temperature	n/k
design temperature [°C]	25
PMV, class II (up)[a]	100 %
PMV, class II (low)[b]	71 %
avg. room temperature [°C][c]	21-22
temp. drift day [K][c]	0.5-1.2
POE[d]	yes
humidity comfort, class II	96 %

MONITORING RESULTS AND BUILDING SIGNATURE (period 5/08 to 11/09)

COOLING	
useful energy [kWh$_{therm}$/m²a]	19.8
final energy [kWh$_{fin}$/m²a]	20.8
primary energy [kWh$_{prim}$/m²a]	n/k
HEATING	
useful energy [kWh$_{therm}$/m²a]	51.9
final energy [kWh$_{fin}$/m²a]	54.5
primary energy [kWh$_{prim}$/m²a]	n/k
VENTILATION	
final energy [kWh$_{fin}$/m²a]	n/k
primary energy [kWh$_{prim}$/m²a]	n/k
LIGHTING	
final energy [kWh$_{fin}$/m²a]	n/k
primary energy [kWh$_{prim}$/m²a]	n/k
APPLIANCES/PLUG LOADS	
final energy [kWh$_{fin}$/m²a]	n/k
TOTAL BUILDING	
final energy [kWh$_{fin}$/m²a]	161.5
primary energy [kWh$_{prim}$/m²a]	215.5
onsite generation of energy	no

(a) upper comfort boundaries, (b) lower comfort boundaries, (c) during occupancy, (d) post-occupancy evaluation

MIXED-MODE COOLING
Athens Greece (38°02', 23°80', 70m)

CLIMATE

annual avg. temp. [°C]	18.9	
month. avg. max temp. [°C]	23.6	
month. avg. min temp. [°C]	15.5	
heating degree days [Kday]	1,388	
cooling degree days [Kday]	271	
design indoor temp. H [°C]	18	
design indoor temp. C [°C]	27	

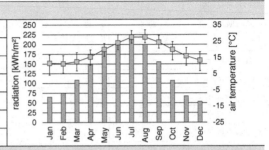

INFORMATION ON BUILDING AND USE

occupancy	office
number of occupants	50
utilization	8am-6pm
completion	1995
refurbishment	-
number of floors	3
total floor area [m²]	1,000
total conditioned area [m²]	600
total volume [m³]	1,296
area-to-volume ratio [m⁻¹]	0.7

Source: Mat Santamouris, Athens

BUILDING ENVELOPE

shading system	exterior fabric blinds, automatic operation, shading factor 0.5
U-values [W/(m²K)]	exterior wall: 0.25 I window: 2.7 I roof: 0.27 I avg. value of building: 0.26
window	2-pane glazing I g-value: 0.7 I area: 53m² I window-façade-ratio: 15-81%

COOLING CONCEPT

environmental heat sink	AA
energy carrier	E
cooling system	a-a-HP, SU
power of system [kW therm]	37
distribution system	air

VENTILATION CONCEPT

operable windows	yes
night-ventilation	f+m
mechanical ventilation	yes
air-change rate [h⁻¹]	no info
dehumidification of air	no
pre-cooling of air	yes

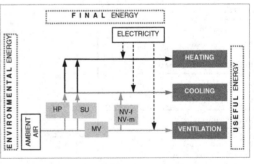

ambient air (AA) I air-to-air heat pump (a-a-HP) I electricity (E) I free (f)mechanical (m) I mechanical ventilation (MV) I night-ventilation (NV) split units (SU)

THERMAL COMFORT PERFORMANCE IN SUMMER

year(s) of monitoring	2008–09
year of evaluation	2008–09
number of rooms	6
interval of measurements	60min
ambient air temperature	at building
design temperature [°C]	27
adaptive, class II (up)[a]	97
adaptive, class II (low)[b]	90
avg. room temperature [°C][c]	23-7
temp. drift day [K][c]	0.8-2.2
POE[d]	yes
humidity comfort, class II	82

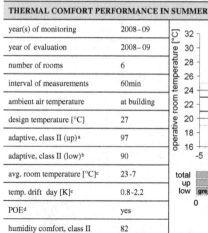

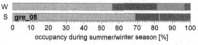

MONITORING RESULTS AND BUILDING SIGNATURE (period 5/08 to 4/09)

COOLING	
useful energy [kWh$_{therm}$/m²a]	19.5
final energy [kWh$_{fin}$/m²a]	18.8
primary energy [kWh$_{prim}$/m²a]	47.0
HEATING	
useful energy [kWh$_{therm}$/m²a]	21.5
final energy [kWh$_{fin}$/m²a]	29.2
primary energy [kWh$_{prim}$/m²a]	73.0
VENTILATION	
final energy [kWh$_{fin}$/m²a]	1.2
primary energy [kWh$_{prim}$/m²a]	3.0
LIGHTING	
final energy [kWh$_{fin}$/m²a]	2.7
primary energy [kWh$_{prim}$/m²a]	6.8
APPLIANCES/PLUG LOADS	
final energy [kWh$_{fin}$/m²a]	27.8
TOTAL BUILDING	
final energy [kWh$_{fin}$/m²a]	51.9
primary energy [kWh$_{prim}$/m²a]	129.8
onsite generation of energy	no

(a) upper comfort boundaries, (b) lower
comfort boundaries, (c) during occupancy, (d)
post-occupancy evaluation

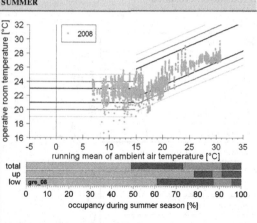

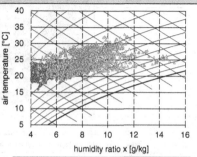

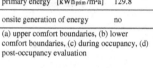

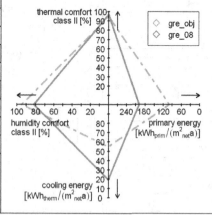

MIXED-MODE COOLING
Bucharest Romania (44°43', 26°9', 55m)

CLIMATE

annual avg. temp. [°C]	10.5
month. avg. max temp. [°C]	17.3
month. avg. min temp. [°C]	7.4
heating degree days [Kday]	2,699
cooling degree days [Kday]	132
design indoor temp. H [°C]	18
design indoor temp. C [°C]	26

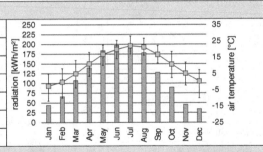

INFORMATION ON BUILDING AND USE

occupancy	office
number of occupants	24
utilization	8am-5pm
completion	1940
refurbishment	-
number of floors	4
total floor area [m²]	634
total conditioned area [m²]	292
total volume [m³]	1.952
area-to-volume ratio [m⁻¹]	0.32

Source: Adrian Ghiaus, Bucharest

BUILDING ENVELOPE

shading system	exterior wood blinds, manual operation, shading factor 0.3
U-values [W/(m²K)]	exterior wall: 0.82 I window: 2.5 I roof: 0.65 I avg. value of building: 0.94
window	2-single pane glazing I g-value: 0.42 I area: 48m² I window-façade-ratio: 30-40%

COOLING CONCEPT

environmental heat sink	AA
energy carrier	E
cooling system	NV-f, SU
power of system [kWtherm]	n/k
distribution system	air

VENTILATION CONCEPT

operable windows	yes
night-ventilation	yes
mechanical ventilation	no
air-change rate [h⁻¹]	n/k
dehumidification of air	no
pre-cooling of air	no

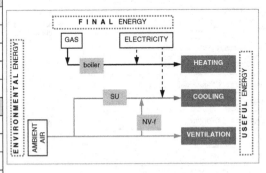

ambient air (AA) I free (f) I electricity (E) I night-ventilation (NV) Isplit unit (SU)

THERMAL COMFORT PERFORMANCE IN SUMMER

year(s) of monitoring	2008 – 09
year of evaluation	2008 – 09
number of rooms	3
interval of measurements	5min
ambient air temperature	public WS[a]
design temperature [°C]	26
adaptive, class II (up)[b]	96 %
adaptive, class II (low)[c]	100 %
avg. room temperature [°C][d]	26-28
temp. drift day [K][d]	1.0-2.5
POE[e]	no
humidity comfort, class II	99 %

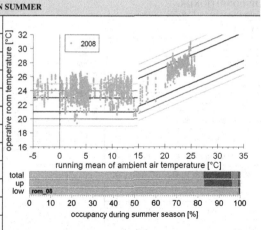

MONITORING RESULTS AND BUILDING SIGNATURE (period 7/08 to 4/09)

COOLING	
useful energy [kWh therm /m²a]	28.3
final energy [kWh fin/m²a]	11.3
primary energy [kWh prim /m²a]	28.3

HEATING	
useful energy [kWh therm /m²a]	n/k
final energy [kWh fin/m²a]	149.6
primary energy [kWh prim /m²a]	164.5

VENTILATION	
final energy [kWh fin /m²a]	1.6
primary energy [kWh prim /m²a]	4.1

LIGHTING	
final energy [kWh fin/m²a]	2.6
primary energy [kWh prim /m²a]	6.5

APPLIANCES/PLUG LOADS	
final energy [kWh fin/m²a]	16.8

TOTAL BUILDING	
final energy [kWh fin/m²a]	165.1
primary energy [kWh prim /m²a]	203.4
onsite generation of energy	no

(a) public weather station, (b) upper comfort
boundaries, (c) lower comfort boundaries, (d)
during occupancy, (e) post-occupancy
evaluation

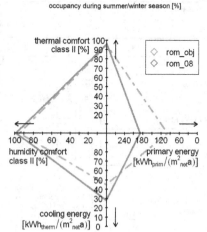

AIR-CONDITIONING
Freiburg Germany (47°99',7°84', 278m)

CLIMATE

annual avg. temp. [°C]	12.9	
month. avg. max temp. [°C]	23.1	
month. avg. min temp. [°C]	3.7	
heating degree days [Kday]	2,771	
cooling degree days [Kday]		
design indoor temp. H [°C]	25	
design indoor temp. C [°C]	20	

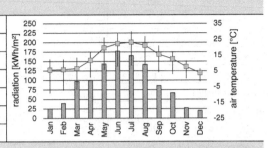

INFORMATION ON BUILDING AND USE

occupancy	Office
number of occupants	400
utilization	8am–5pm
completion	2000
refurbishment	-
number of floors	4
total floor area [m²]	6,880
total conditioned area [m²]	4,131
total volume [m³]	20,640
area-to-volume ratio [m⁻¹]	n/k

Source: Fototeam Vollmer, Freiburg

BUILDING ENVELOPE

shading system	exterior venetian blinds, automatic and manual operation, shading factor 0.2
U-values [W/(m²K)]	exterior wall: 0.25 I window: 1.3
window	heat protection glazing I g-value 0.58 I window-façade-ratio: 30–40%

COOLING CONCEPT

environmental heat sink	AA,GR
energy carrier	E, DC
cooling system	rev. HP
power of system [kW$_{therm}$]	110
distribution system	CP-w

VENTILATION CONCEPT

operable windows	no
night-ventilation	no
mechanical ventilation	yes
air-change rate [h⁻¹]	1.6
dehumidification of air	yes
pre-cooling of air	yes

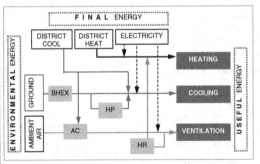

ambient air (AA) I air-conditioning (AC) I borehole heat exchanger (BHEX) Iwater-driven ceiling-suspended cooling panels (CP-w) I district cooling (DC) I district heating (DH) I electricity (E) I heat pump (HP) heat recovery (HR)

THERMAL COMFORT PERFORMANCE IN SUMMER

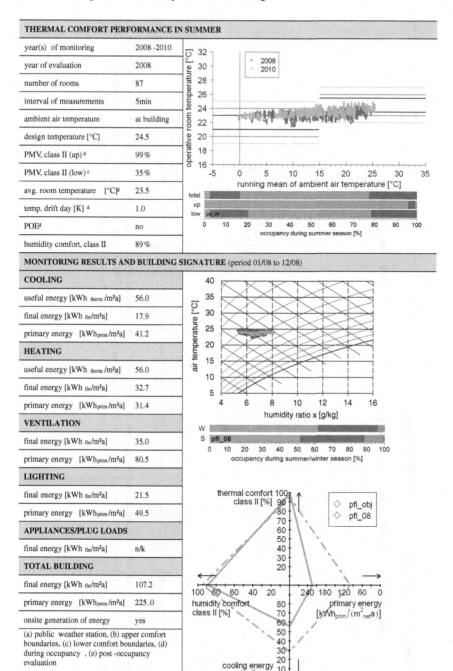

year(s) of monitoring	2008 -2010
year of evaluation	2008
number of rooms	87
interval of measurements	5min
ambient air temperature	at building
design temperature [°C]	24.5
PMV, class II (up) [b]	99 %
PMV, class II (low) [c]	35 %
avg. room temperature [°C][d]	23.5
temp. drift day [K] [d]	1.0
POE[d]	no
humidity comfort, class II	89 %

MONITORING RESULTS AND BUILDING SIGNATURE (period 01/08 to 12/08)

COOLING	
useful energy [kWh therm /m²a]	56.0
final energy [kWh fin/m²a]	17.9
primary energy [kWhprim/m²a]	41.2

HEATING	
useful energy [kWh therm /m²a]	56.0
final energy [kWh fin/m²a]	32.7
primary energy [kWhprim/m²a]	31.4

VENTILATION	
final energy [kWh fin/m²a]	35.0
primary energy [kWhprim/m²a]	80.5

LIGHTING	
final energy [kWh fin/m²a]	21.5
primary energy [kWhprim/m²a]	49.5

APPLIANCES/PLUG LOADS	
final energy [kWh fin/m²a]	n/k

TOTAL BUILDING	
final energy [kWh fin/m²a]	107.2
primary energy [kWhprim/m²a]	225.0
onsite generation of energy	yes

(a) public weather station, (b) upper comfort
boundaries, (c) lower comfort boundaries, (d)
during occupancy , (e) post -occupancy
evaluation

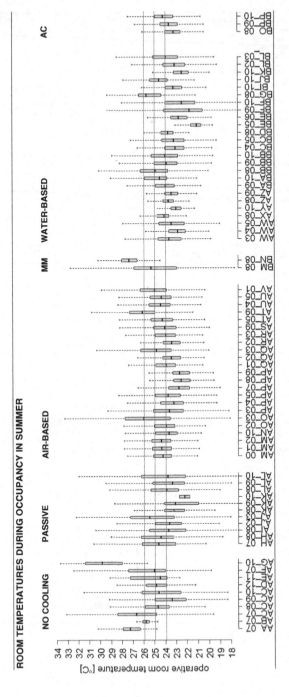

Fig. 6.3 Operative room temperatures during occupancy in summer [°C], illustrated as boxplots. Buildings are grouped in accordance with the building categories introduced. Further given values are the median (*solid line*), the 75th and the 25th percentiles (*grey rectangle*). Building categories: no cooling, passive cooling, air-based mechanical cooling, water-based mechanical cooling, mixed-mode cooling, and air-conditioning

6.3 Cross-Comparison of Thermal-Comfort Performance in Summer for 42 Office Buildings in Europe

This chapter presents a cross-comparison of the thermal-comfort performance of European buildings in summer. Again, a building is allocated to the particular comfort class if the room temperatures remain within the defined comfort boundaries, i.e., do not exceed the upper comfort boundaries (for the evaluation methodology, see Chap. 5).

Figure 6.3 illustrates the range of operative room temperatures during the time of occupancy in summer for the individual buildings. Monitoring results are presented as boxplots indicating 50 % of the temperature measurements as well as maximum and minimum occurrences. Obviously, temperature values are scattered over a wide range, considering both the individual building as well as all buildings within one category. This depends on the prevailing ambient air temperature, the building physics, the use of the buildings, the behavior of the occupant, and the cooling system.

Figure 6.4 portrays measured operative room temperatures during occupancy over the running mean ambient air temperature as comfort figures in accordance with the EN 15251 standard. Results are given for one building chosen from each category. Furthermore, hourly operative room and ambient air temperatures are given for a hot summer week.

Finally, thermal-comfort results are presented as a "thermal-comfort footprint" for all buildings (Fig. 6.5). Results on thermal comfort consider the exceedance of the upper comfort boundaries only for both the PMV and the adaptive comfort approach.

Buildings without cooling. Thermal comfort in buildings without cooling is mainly evaluated based on short-term monitoring campaigns over two to four weeks during hot summer periods (Table 6.1). Obviously, the resulting operative room temperatures are high; mean values are above 25 °C and maximum values are in the range of 29–34 °C. During the monitoring period, only comfort class III—following the adaptive model—is achieved during 65–95 % of the occupancy. One building reveals a pronouncedly poor performance since it was built the 1960s and has not been retrofitted. In general, the buildings without any cooling measures are sensitive to the prevailing ambient conditions. Operative room temperatures rise considerably during warmer periods, resulting in temperature drifts of up to 5 K per day. Outside the occupancy, i.e., during nighttime hours, room temperatures fall just slightly by 1–2 K. Interestingly, relatively high room temperatures are reached even at moderate running mean ambient air temperatures of 18–22 °C.

Buildings with passive cooling. Passive cooling includes all measures to reduce solar and internal heat loads and to store the remaining heat gains in the building mass in order to dissipate them during nighttime through free ventilation. Air-exchange rates vary from day to day and can be different in each office room. The effective exchange rates do often not exceed values of 2 ACH per hour in the

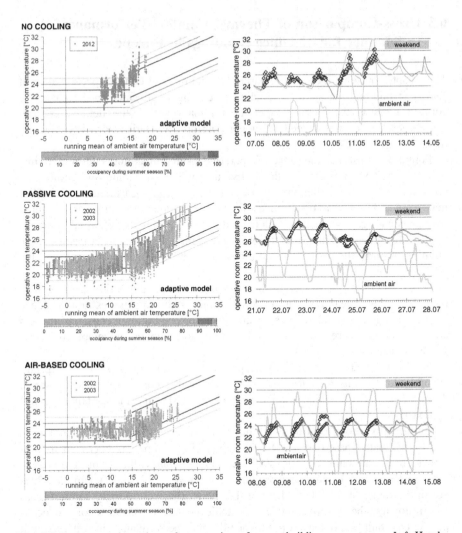

Fig. 6.4 Results on thermal comfort are given for one building per category *Left* Hourly operative room temperatures [°C] during occupancy are portrayed above the running mean ambient air temperature [°C]. Results are given for up to two operation years. Adaptive-comfort approach for buildings without, with passive, air-based and mixed-mode cooling. PMV-comfort approach for buildings with water-based cooling and air-conditioning *Right* Hourly room temperatures for two reference rooms and the prevailing ambient air temperature is given for one hot summer week. Time of daily occupancy is indicated by markers

northern summer climate zones and rarely reach values above 1.8 ACH per hour in the southern ones. However, these relatively straightforward measures for protection against summer overheating improve the comfort performance of the buildings significantly as compared to buildings without cooling measures.

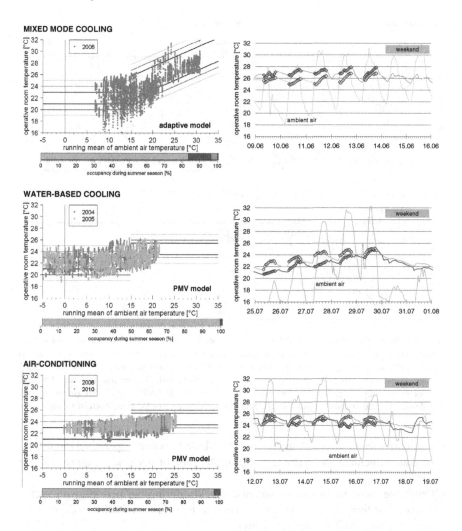

Fig. 6.4 continued

- Relatively high operative room temperatures can be expected during the summer season due to high interior heat gains and often densely occupied buildings. Most of the buildings show an exceedance of comfort class II in accordance with the adaptive-comfort model during 2–10 % of the time of occupancy. During summer heat waves, the indoor temperatures rise significantly above 28 °C. However, comfort requirements of class III are violated only during few hours in summer when maximum temperatures reach values of 28–30 °C. The middle 50 % of recorded temperature values during summer occupancy are within a range of 23–26 °C.

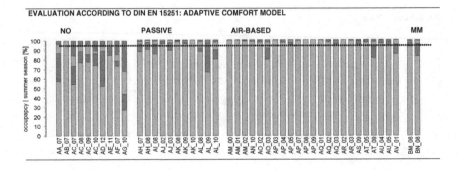

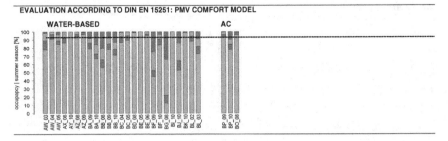

Fig. 6.5 Thermal-comfort footprint for all demonstration buildings. This figure shows the time of occupancy [%] during summer season when the comfort requirements of classes I, II, and III are achieved in accordance with the adaptive- or PMV-comfort approach of the EN 15251 standard. Only exceedance of *upper* comfort limits is considered. *Dashed line* indicates 95 % of time of occupancy. Legend: comfort class I (*light green*), II (*dark green*), III (*orange*), and outside the defined comfort classes (*red*). Squares for classes I to IV from bottom to top

- Compared to buildings with air- or water-based cooling, operative room temperatures fluctuate more strongly during the time of occupancy, up to 3 K per day on average.
- Operative room temperatures increase quickly during hot periods with higher ambient air temperatures. However, cooling by night-ventilation reduces temperatures by 2 to 3 K during the night.
- Passive cooling is applicable in northern Europe because relatively high solar heat gains—due to long sunshine hours with the sun at a lower angle—can be dissipated efficiently by cool outdoor air during nighttime, provided that the building is planned explicitly for this concept under consideration of the particular local conditions. Nevertheless, mechanical night-ventilation improves the controllability and heat dissipation during midsummer periods.
- Surprisingly, room-temperature measurements clearly violate the lower comfort boundaries, in particular during summer periods with moderate ambient air temperatures at a running outdoor mean of 15–20 °C. Most of the buildings employ free night-ventilation by means of open windows and ventilation slats, which does not enable influencing the room temperature. However, room temperatures rarely fall below 21 and 20 °C.

Buildings with air-based mechanical cooling

- A ventilation system ensures a good indoor air quality during daytime and can be used for mechanical night-ventilation. The ventilation unit is usually operated when the room temperature exceeds 21 °C and the outdoor temperature is at least 2 K below room temperature.
- Experience with buildings cooled by mechanical night-ventilation shows that pleasant room temperatures are reached in summer and that in general, occupants rate the indoor climate positively. During 95 % of the occupancy, the buildings with air-based mechanical cooling comply with comfort class II, considering the adaptive approach. Just two of the buildings studied only achieved comfort class III.
- Buildings that employ mechanical night-ventilation by means of an exhaust or supply-and-exhaust system allow for a fixed air-change rate of between 2 and 4 ACH. Therefore, the cooling capacity is higher and less dependent on the prevailing wind situation and differences between indoor and outdoor temperatures.
- Rising ambient air temperatures decrease the cooling potential of night-ventilation and demand elevated air-change rates for a mechanical ventilation system. Besides, persistent heat waves in central and southern climates with elevated outdoor temperatures at night prevent effective cooling of the thermal mass of the building. Then, structural measures and mechanically assisted night-ventilation is often not sufficient to limit room temperatures to 28 °C during daytime. For example: although the building AO in south-west Germany—which uses night-ventilation by means of an exhaust ventilation system only—performed surprisingly well in 2002 (class I), a persistent heat wave in 2003 resulted in an extensive violation of comfort class II, considering the adaptive-comfort approach. On the contrary, the building AW, using water-based cooling by TABS in combination with the ground as a heat sink, could be assigned to comfort class II in summer 2003, corresponding to the PMV-comfort approach. The night-ventilated buildings AQ and AR in Germany provided good thermal comfort during the hot summer of 2003. This reduced cooling demand attributes to the high quality of the building envelope and the stringent load-reduction strategy (day-lighting concept, solar-shading system, energy-efficient office equipment). Nevertheless, resulting operative room temperatures in the buildings studied are significantly below the prevailing ambient air temperature.
- Evidently, air-based building systems are not as effective as water-based thermo-active ones, in particular during prolonged periods with high ambient air temperatures. Taking the extreme summer conditions of 2003 as an indicator for a global warming scenario, thermal comfort in compliance with class II of EN 15251:2007-08 cannot be ensured in buildings that employ night-ventilation only.
- TABS are essentially unaffected by these disadvantages and thus present an effective concept for conditioning buildings in Central European climates, even

in very hot summers, given a building with a high-quality envelope, solar-shading devices, and reduced internal loads.

Buildings with mechanical water-based cooling

- Compared to buildings with air-based cooling, operative room temperatures are lower. The middle 50 % of the measurements recorded are within a range of 22.5 and 25.5 °C. Furthermore, temperatures fluctuate less during the day, and maximum temperature values are above 27 °C in five buildings only.
- Except for two buildings, water-based mechanical cooling by means of TABS and environmental energy provides good thermal comfort during summer. By applying the adaptive comfort approach, most buildings meet the upper comfort requirements of class II.
- Considering the PMV-comfort approach of EN 15251, only eight out of 16 buildings meet the strict requirements for comfort class II over the course of the summer. The reason for an insufficient thermal comfort according to the PMV approach in those buildings can often be attributed to an improperly sized cooling system (e.g., cooling capacity of heat sink not sufficient at buildings BA and BK) or to inadequate control and operation algorithms (e.g., building BE during the first year of operation). In some cases, the plant system was designed to provide limited cooling only without claiming to guarantee the stringent temperature setpoints of the PMV approach.
- The restricted cooling capacity of concrete-core conditioning systems requires a consequent reduction of interior and solar-heat gains through a holistic design process considering the building physics, architecture, HVAC systems, and the use of the buildings. The combination of concrete-core conditioning with a controllable and thermally fast-responding auxiliary cooling system (TABS near surface or suspended systems) is reasonable—in case of a higher cooling demand—for office spaces with changing conditions of use or in areas with higher comfort requirements, e.g., meeting rooms. The additional cooling system should also be operated at the same temperature level as the concrete-core conditioning system. This allows for the use of the same distribution system and therefore for saving investment costs.
- Obvious is the frequent violation of the lower comfort boundaries in summer, even during periods with higher ambient air temperatures. This can be observed in many buildings studied. Although users rate these temperatures as slightly cool, their satisfaction is high in general (see Chap. 4). Temperatures below the lower comfort boundaries can be avoided through system controls based on outside and room temperatures. In addition, the required thermal cooling energy is reduced as well.
- In northern European buildings, thermal-comfort requirements of class II according to the PMV approach are met at more than 95 % of the occupancy. That means that room temperatures are usually below 26 °C and are subject to only small daily fluctuations (of between 0.8 and 1.5 K).

- In northern European climates, water-based cooling by means of environmental heat sinks or even active cooling with compression chillers is only required for buildings with very high comfort requirements (class I) or limited user influence (dress codes, sealed windows). As against, in southern European summer climates, a relatively high cooling capacity must be provided to dissipate strongly fluctuating cooling loads. Since the temperature difference between the ambient air and the indoor comfort temperature is low, an active cooling system is often required to meet the comfort requirements. Water-based cooling concepts are generally suitable in all climate zones of Europe. However, thermally slow responding systems such as concrete-core conditioning systems reach the cooling capacity limits during periods of high and fluctuating head loads under southern European climate conditions.

Buildings with mixed-mode cooling

- The two buildings with mixed-mode cooling in Greece and Romania are primarily and—as far as possible—cooled with night-ventilation. In periods of an increased cooling demand, office rooms are additionally air-conditioned via decentralized split units. The monitoring in the buildings does not allow for distinguishing the comfort analysis in accordance with operating modes, i.e., night-ventilation and/or active cooling. Even during periods of high outdoor temperatures (running mean ambient air temperature between 26 and 32 °C), the measured room temperatures stay within the required limits of comfort class II, following the adaptive approach.

Buildings with air-conditioning

- Air-conditioning systems operate reliably in all climate zones, providing a sufficiently high cooling capacity to meet the high standards of comfort, even at times of high outdoor temperatures. The two German buildings as examples for full air-conditioning provided excellent thermal comfort during summer, considering the upper boundaries of the PMV-comfort approach. Room temperatures are usually between 24 and 26 °C and are subject only to very small daily fluctuations. An air-conditioning system allows for well-defined and narrow temperature ranges during the time of occupancy and the dehumidification of the supplied air. However, monitoring results show a tremendous violation of the lower comfort boundaries as well, which might result in dissatisfied occupants feeling too cold.
- While northern and central European buildings can provide satisfactory thermal comfort without air-conditioning and meet requirements for indoor humidity, the room temperature and the relative humidity of moist, warm days are higher in southern climates, i.e., the relative humidity is in the range of the upper limit of the standard. Then, dehumidification of the supply air appears—at least temporarily—necessary. In the case of using an air-conditioning system, the dissipation of sensible cooling loads should be realized with a water-based cooling system in order to provide cooling energy with a higher efficiency by

using—if possible—environmental heat sinks such as ambient air or the surface-near ground. Latent cooling loads are dissipated by the air-conditioning units.

6.4 Humidity Performance in Summer

In accordance with EN 15251:2007-08, a humidification of indoor air is usually not needed. Humidity has only a small effect on thermal sensation and perceived air quality in the rooms of sedentary occupancy; however, long-term high humidity indoors will cause microbial growth, and very low humidity (<15–20 %) causes dryness and irritation of eyes and airways. Requirements for humidity influence the design of dehumidifying (cooling load) and humidifying systems and therefore energy consumption. The criteria depend partly on the requirements for thermal comfort and indoor air quality and partly on the physical requirements of the building (condensation, mold, etc.) Design-limit values for relative humidity concerning dehumidification are 50 % for class I, 60 % for class II, and 70 % for class III in accordance with the DIN ISO 7730:2005 (ISO 7730) standard. Design-limit values for relative humidity concerning humidification are 30 % for class I, 25 % for class II, and 20 % for class III. The values apply to the summer and winter periods, i.e., there is no dependency of them on the prevailing relative humidity of the outdoor air.

The building profiles given in this chapter present the monitoring results of indoor humidity conditions in the buildings in accordance with EN 15251:2007-08. Main results are:

- The northern and mid-European buildings provided good humidity comfort with respect to the guideline. Class II was achieved in the German, Danish, Finish, and Romanian buildings.
- The buildings in southern and south-western Europe (France, Greece, Italy) showed higher values for relative humidity and, therefore, violated classes I and II. Class III was achieved only during 75–80 % of the time of occupancy.

6.5 Conclusion: Cross-Comparison of Thermal Comfort

- The mean room temperature during the time of occupancy in summer is 25 °C for the buildings with passive and air-based mechanical cooling. Considering the water-based mechanical and air-conditioned buildings, the mean room temperature is slightly lower at 23.5 °C.
- Maximum temperature occurrences are about 1–2 K higher in the buildings with air-based mechanical cooling than in those with water-based and air-conditioned cooling.

- Each category contains buildings that exceed significantly the mean operative temperature levels of the category by up to 3 K.
- Obviously, the occurring range of operative room temperatures in the buildings with passive and air-based mechanical cooling is markedly wider than in the other buildings.
- Considering the three buildings in southern and south-eastern Europe (Italy, Greece, and Romania), room temperatures are elevated significantly in comparison to the northern European and German buildings. (Cooling concepts for different European climates will be further studied in Chap. 7).
- The German building with full air-conditioning (cooling and dehumidification) has the narrowest temperature band during occupancy, that is 22.5–23.5 °C.
- Surprisingly, the room temperatures of the buildings with water-based cooling are scattered noticeably around the mean temperature of 23.5 °C. Some buildings have relatively elevated temperature ranges. On the contrary, some buildings are mainly below the average temperature range of their respective category. Low-energy buildings with heavy-weight constructions reduce and cover the energy demand mainly with passive technologies and environmental heat sources and sinks. Due to their thermal mass, they behave indulgently towards exterior and interior changes, e.g., higher ambient air temperatures or higher internal loads due to open solar shading or open windows. This means that the heavy-weight building construction buffers the rise of the operative room temperature for a certain period, avoiding overheated offices and rooms.
- Monitoring results indicate that the buildings with radiant cooling and environmental-energy systems with lower cooling capacities are nevertheless sensitive towards the applied control and operation algorithms as well as occupant behavior. Therefore, room temperature cannot always be kept within a narrow range.
- Evidently, the comfort performance is not strongly affected by the type of environmental heat sink employed (i.e., using the ground, groundwater, ambient air), provided that the heat sink is adequately dimensioned and well-operated.
- An unexpected result is the wide range of comfort ratings within a given building, e.g., some monitored rooms do not violate the comfort boundaries at all, whereas others reveal a significant exceedance. This discrepancy within a building is mainly affected by the orientation of the rooms (Fig. 6.4), the presence of the occupants and their behavior in terms of opening windows and using solar shading, which is not monitored in this investigation.
- The use of the building, that is, the user profile and the equipment of office space, as well as the requirements for thermal comfort must be clearly defined in the planning phase of a building and have to be considered in the building's operation. Its use and user behavior have a significant impact on room comfort and energy consumption for cooling the built environment. Users should be informed about the building and energy concept and should receive comprehensible instructions on how to behave in order to ensure a high level of interior comfort with low energy consumption and costs.
- Convincing building concepts are characterized by the fact that users are enabled to influence their interior surroundings because research studies reveal

that user satisfaction with the interior comfort is affected and increased effectively by the possibility to exert influence on room conditions.
• The present results also suggest that the expectations of users on room and comfort conditions have a considerable impact on their perception and satisfaction: in buildings with night-ventilation, users expect higher room temperatures and thus accept these. On the contrary, users have higher expectations on indoor comfort in buildings with water-based cooling concepts and, therefore, are less satisfied with higher room temperatures.

References

Brager G, Borgeson S, Lee Y (2007) Summary report: control strategies for mixed-mode buildings. Technical report, University of California, Berkeley
DIN EN 60751:2009–05 Industrial platinum resistance thermometers and platinum temperature sensors (IEC 60751:2008). Beuth, Berlin

Chapter 7
Application of Cooling Concepts to European Office Buildings

Abstract A simulation study investigates the potential of different ventilation and cooling strategies with regard to energy efficiency and thermal comfort in different European climates. The results demonstrate a high potential for night-ventilation strategies in the northern European climate with its low ambient air temperatures. In the mid-European climate, water-based low-energy cooling technologies based on radiant cooling make use of the cool ground in summer. Active cooling provides good thermal comfort in the southern European climate, with high and fluctuating cooling loads.

A simulation study investigates the potential of different ventilation and cooling strategies with regard to energy efficiency and thermal comfort in different European climates. The results demonstrate a high potential for night-ventilation strategies in the northern European climate with its low ambient air temperatures. In the mid-European climate, water-based low-energy cooling technologies based on radiant cooling make use of the cool ground in summer. Active cooling provides good thermal comfort in the southern European climate, with high and fluctuating cooling loads.

7.1 Simulation Study of Cooling Concepts

Building Model. The simulation model is a new three-floor, two-wing office building and contains two office rows with a dimension of 5.2 m in length, 3.9 m in width, and 3.0 m in height for each office, which are separated by a corridor (width: 2.6 m). The building is simulated in North–South and East–West orientation.

The simulation model represents a typical European office building with an area-to-volume ratio of 0.4 $m^2_{ext.surface}/m^3_{int.volume}$ and a window ratio of 0.32 $m^2_{window}/m^2_{ext.wall}$ (Fig 7.1). The building's physical properties meet the EPBD requirements:

D. E. Kalz and J. Pfafferott, *Thermal Comfort and Energy-Efficient Cooling of Nonresidential Buildings*, SpringerBriefs in Applied Sciences and Technology, DOI: 10.1007/978-3-319-04582-5_7, © The Author(s) 2014

Fig. 7.1 Building simulation model: typical European office building for northern, mid- and southern European countries

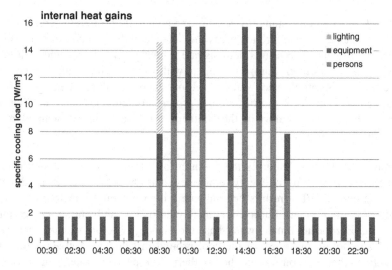

Fig. 7.2 Internal heat gains during working days in summer period: the internal heat gains from artificial lighting differs from month to month and with latitude

- External walls, baseplate, and ceiling: $U_{mean} = 0.24$ W/m^{2}K incl. thermal heat bridges.
- Windows: $U_w = 1.0$ W/m^{2}K and $g = 0.58$.
- Solar shading: external Venetian blinds ($F_c = 0.06$, $F_c = 0.2$, considering nonoptimal closing) are closed semiautomatically once the solar radiation on the façade exceeds 200 W/m^2.

The offices are occupied from 8 a.m. to 6 p.m. (UTC) during workdays. The daily internal heat gains are 156 Wh/m^{2}d, with a standardized load profile as shown in Fig. 7.2.

Plant Model and Cooling Concepts. Five different cooling concepts are applied in order to cool the office building (Fig. 7.3). All of them allow for free ventilation by opening windows. Four concepts employ exhaust fans in order to provide a

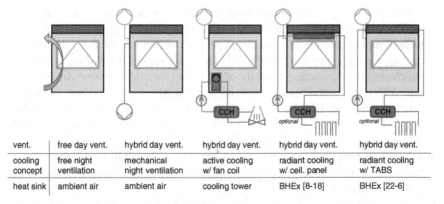

vent.	free day vent.	hybrid day vent.	hybrid day vent.	hybrid day vent.	hybrid day vent.
cooling concept	free night ventilation	mechanical night ventilation	active cooling w/ fan coil	radiant cooling w/ ceil. panel	radiant cooling w/ TABS
heat sink	ambient air	ambient air	cooling tower	BHEx [8-18]	BHEx [22-6]

Fig. 7.3 Five different cooling concepts: passive cooling, night-ventilation, active cooling with compression chiller, and water-based low-energy cooling (with compression chiller if needed to meet the cooling load). *Green* ventilation, *blue* cooling, and *gray* heat sink

minimum air-change rate of 40 m³/h per person. Although in actual projects, an exhaust-and-supply air system may be applied in order to pre-heat the air in winter by means of a heat-recovery system and/or to dehumidify the supply air in summer, the simulation study considers only an exhaust-air system for better comparison to the sensible cooling capacities of low-energy cooling concepts.

- Passive cooling refers to techniques used to prevent and modulate heat gains. The reduced cooling loads can be dissipated with free ventilation only. The building has external shading in order to avoid overheating. A bare concrete ceiling modulates the internal and solar heat gains. Open windows during the night enable an increased single-side- and cross-ventilation.
- In the investigated office building, the air-change rates differ from day-to-day and from location to location. It often exceeds 2 h^{-1} in the cooler summer climates, while warmer summer nights in southern Europe allow for maximum air-change rates of 1.8 h^{-1}.
- An exhaust-ventilation system can also be used for mechanical night-ventilation and provides an air-change rate whenever the room temperature exceeds 21 °C, with a minimum temperature difference of 2 K between inside and outside.
- A fan-coil unit is simulated as a reference system. It provides sensible cooling to meet the comfort requirements during the time of occupancy. A compression chiller provides cold water with a supply temperature of 13 °C. The design return-temperature is 18 °C. A cooling tower is used for recooling. The maximum cooling capacity is limited to 1.8 kW or 90 W/m², respectively.
- The coefficient of performance (COP) decreases from North to South due to increasing ambient air temperatures during the time of operation. The mean COP_{mean} for this timeframe decreases from 3.1 kW$_{cooling}$/kW$_{el}$ in Stockholm to 2.4 kW$_{cooling}$/kW$_{el}$ in Palermo.

- A radiant cooling ceiling panel is operated during the time of occupancy. Its cooling capacity is a function of the difference between mean cold water and room temperature. For a typical temperature difference of 8 K, the specific cooling capacity is approx. 100 W/m^2. The panel covers 70 % of the office area, which results in a cooling capacity of 70 W/m^2 for a temperature difference of 8 K. The actual maximum cooling capacity in Milano is 77 W/m^2 for a temperature difference of 9 K. The supply temperature is controlled by the equation T_{supply} [°C] = 18 °C + 0.35 (18 °C—$T_{ambient}$ [°C]), with a minimum supply temperature of 16 °C to avoid condensation.
- A borehole heat exchanger is used as heat sink. The undisturbed ground temperature in summer is calculated for each climate zone and increases from 6.3 °C in the North to 19.6 °C in the South.
- If the return temperature from the borehole heat exchanger exceeds the set temperature, an optional compression chiller will provide additional cooling. As the borehole heat exchanger in Stuttgart provides cool water during the whole summer, the seasonal energy efficiency ratio (SEER) is 14 kWh$_{therm}$/kWh$_{el}$. In Rome, active cooling is needed and, hence, the SEER is 3.4 kWh$_{therm}$/kWh$_{el}$ only.
- A concrete-core conditioning cools the whole ceiling during the night. Due to high thermal inertia, the mean cooling capacity of approx. 40 W/m^2 is provided throughout the day. This results in a considerable fluctuation of room temperatures during the time of occupancy.
- The control strategy is similar to the operation of the radiant panel, but with night time operation. The seasonal energy efficiency ratio SEER is 14 kWh$_{therm}$/kWh$_{el}$ in Stuttgart. In Rome, the SEER is 3.8 kWh$_{therm}$/kWh$_{el}$, due to higher supply temperatures than for the operation of the radiant cooling ceiling panels.

Investment Costs. The investment costs are calculated for the typical office building shown in Fig. 7.1. These cost estimations are based on an analysis of realized HVAC concepts in Germany (Voss et al. 2006; Voss and Pfafferott 2007):

- Passive cooling: 20 €/m^2. Ventilation slats and enlarged openings for a lower pressure drop.
- Mechanical night-ventilation: 32 €/m^2. Ventilation slats, exhaust ventilator, and ducting. Control system.
- Fan coil: 85 €/m^2. Ventilation slats, exhaust ventilator, and ducting. Compression chiller with cooling tower, fan-coil units, and cold-water piping. Control system.
- Radiant cooling ceiling panel: 138 €/m^2. Ventilation slats, exhaust ventilator, and ducting. Compression chiller with borehole heat exchanger, suspended radiant cooling and heating panel, and piping. Control system.
- Thermo-active building system: 117 €/m^2. Ventilation slats, exhaust ventilator, and ducting. Compression chiller with borehole heat exchanger, concrete-core conditioning, and piping. Control system.

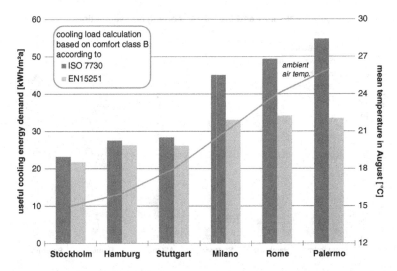

Fig. 7.4 Useful cooling-energy demand: the cooling-energy demand increases from North to South. The static-comfort model (ISO 7730) results in a higher cooling-energy demand than the adaptive comfort model (EN 15251)

Climate. The simulation study was carried out for six different European climate zones. Each climate zone is defined by the mean ambient air temperature in August and characterized by a meteorological reference station. The summer temperatures stay below 16 °C in Stockholm, between 16 and 18 °C in Hamburg, 18 and 20 °C in Stuttgart, 20 and 22 °C in Milano, 22 and 24 °C in Rome, and exceed 24 °C in Palermo. Note: these summer climate zones correspond reasonably with the USDA Hardiness Zones from 5 (in the North) to 10 (in the South).

Building and Plant Simulation. The coupled building and plant simulation is run for the summer period from May to September. The cooling load is calculated for both the static-comfort model in accordance with ISO 7730 (2005) and the adaptive model in accordance with EN 15251 (2007). The cooling capacity and the final energy use for cooling are calculated only for the operative room temperature in accordance with the adaptive comfort model.

7.2 Simulation Results and Conclusions

The cooling load [W/m^2] increases from North to South, mainly due to higher temperatures and to a lesser extent due to higher solar heat gains. Beyond that, Fig. 7.4 shows that the useful cooling-energy demand [kWh/m^{2}a] is also a function of the comfort criteria to be met. If the daily mean temperature in summer is considerably lower than the room temperature, the comfort temperature does not differ significantly. Hence, the useful cooling-energy demand is similar for both comfort models in northern European climates but differs in southern Europe. In

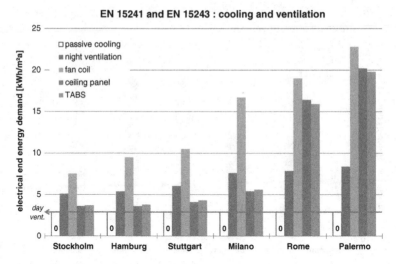

Fig. 7.5 Final energy demand for five cooling concepts in six locations: the cooling concepts do not necessarily meet the comfort requirements

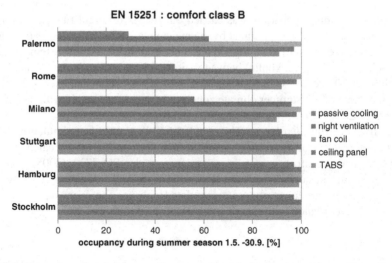

Fig. 7.6 Thermal comfort for five cooling concepts in six locations: passive cooling, air- and water-based low-energy cooling (if needed with compression chiller to meet the cooling load), and active cooling with compression chiller

Hamburg, the useful cooling-energy demand is 27.6 and 26.3 kWh/m²a, respectively. Compared to this, it is higher in Rome and differs considerably from 34.1 to 49.4 kWh/m²a for the two comfort models. These findings correspond to the results of the COMMONCENSE research project (Santamouris and Sfakianaki 2009).

Table 7.1 Application of cooling concepts to European climates: results of a simulation study with a typical office building in different European climates

	Passive cooling[a]	Air-based mechanical cooling	Water-based mechanical cooling		
Ventilation during occupancy					
	free	1.3 ACH	1.3 ACH	1.3 ACH	1.3 ACH
Ventilation at nighttime					
	free	4 ACH	no	no	no
Active cooling					
	no	no	fan coil [b]	ceiling panel [c]	TABS [c,d]
Investment costs (€/m²)					
	20	32	85	138	117
Application in European climates					
Stockholm (hr)	++	+	−	+	+
Hamburg (hr)	+	++	−	+	+
Stuttgart (hr + dh)	−	+	−	++	++
Milano (hr + dh)	x	−	−	++	++
Rome (dh)	x	x	+	++	+
Palermo (dh)	x	x	++	+	−

Cooling concepts are rated in accordance with the thermal comfort achieved during occupancy, the cooling energy used, and the energy efficiency of the system. The specific investment costs are considered as an additional criterion

Legend: (++) preferential concept, (+) good concept, (+/++) good but comparatively expensive concept, (−) unfavorable concept, (x) concept not applicable to the respective climate, (ACH) air-change rate per hour

[a] if applicable with consideration to noise, security, etc.,

[b] can also be used for heating in winter

[c] should be used with ground-coupled heat pump for heating in winter

[d] for new buildings only Ventilation concept in real building design: hr supply-and-exhaust air ventilation with heat recovery in winter, dh supply-and-exhaust air ventilation with dehumidification in summer

The final energy use for cooling and ventilation is calculated in accordance with EN 15241 and EN 15243 for the comfort temperature, in accordance with the adaptive model in EN 15251. There is no energy demand for passive cooling, and for the mechanical ventilation during the time of occupancy, it is 2.9 kWh/m²a for all locations.

Figure 7.5 shows that the energy demand for different cooling concepts does not differ considerably in northern climates (Stockholm and Hamburg). For mechanical night-ventilation, it increases slightly from Mid-European climates (Stuttgart and Milano) to southern European ones (Rome and Palermo), since this concept reaches its capacity limit and cannot provide thermal comfort in hot summer climates. The energy demand for water-based cooling increases significantly from Mid- to southern European climates since the compression chiller has to provide additional cooling. In hot summer climates, the energy demand for active cooling through fan coils is insignificantly higher than for water-based

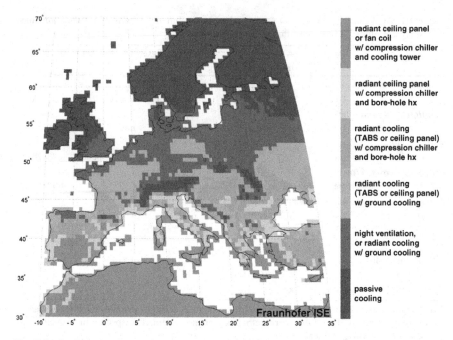

Fig. 7.7 Cooling concepts at six locations: passive cooling, mechanical night-ventilation, water-based low-energy cooling with borehole heat exchanger as heat sink (and optional compression chiller—if needed—in order to meet the cooling load), and active cooling with compression chiller and cooling tower as heat sink

radiant cooling; however, these quick-responding cooling concepts do not allow for peak-load shifting.

Figure 7.6 clearly indicates the limits of each concept with regard to thermal comfort:

- Passive cooling and night-ventilation concepts cannot provide thermal comfort for typical office buildings in all European climates.
- Water-based low-energy cooling can be successfully applied to office buildings in all climate zones and may be operated with additional active cooling.
- A fan coil provides thermal comfort more or less independently from prevailing weather conditions. Therefore, fan coils might be an acceptable solution for buildings in hot summer regions. In contrast to a VRF system or individual room air conditioners, the central compression chiller for cold-water supply allows for some load management.

Figure 7.5 does not consider whether or not a specific cooling concept can provide thermal comfort. Furthermore, Fig. 7.6 does not consider the energy demand needed to provide thermal comfort. Table 7.1 combines these results and classifies the cooling concepts with regard to both aspects. As some concepts are comparable in terms of energy efficiency and achievement of thermal comfort, the investment costs for these cooling concepts are used as third decision parameter.

Figure 7.7 gives an overview of preferential cooling concepts in different European regions.

- In northern European climates, high solar heat gains due to long-standing sunshine can efficiently be dissipated by cool ambient air. In some situations, mechanical night-ventilation is recommended.
- In Mid-European climates, water-based low-energy cooling makes use of the cool ground in summer. If additional active cooling is needed, thermo-active building systems (TABS) with high thermal inertia allow for peak-load shifting.
- In southern European climates, high cooling loads demand concepts with high cooling capacities. Since the temperature differences between ambient heat sinks and comfort temperature are too low, active cooling is needed in order to provide thermal comfort.

References

DIN EN ISO 7730:2005 (2005) Ergonomics of the thermal environment—Analytical determination and interpretation of thermal comfort using calculation of the PMV and PPD indices and local thermal comfort criteria. Beuth, Berlin

DIN EN 15251:2007–08 (2007) Criteria for the indoor environment. Beuth, Berlin

Santamouris M, Sfakianaki K (2009) Predicted energy consumption of major types of buildings in European climates based on the application of EN 15251. Report for EIE/07/190/SI2.467619, COMMONCENSE, Comfort monitoring for CEN standard EN 15251 linked to EPBD. www.commoncense.info. Accessed Dec 2013

Voss K, Löhnert G, Herkel S, Wagner A, Wambsganß M (2006) Bürogebäude mit Zukunft. SOLARPRAXIS Verlag, Köln

Voss K, Pfafferott J (2007) Energieeinsparung contra Behaglichkeit?. Bundesamt für Bauwesen und Raumordnung, Bonn

Chapter 8
Thermal Comfort and Energy-Efficient Cooling

Abstract This chapter summarizes key findings and clarifies success factors for low-energy cooling. Furthermore, a holistic approach is proposed for the evaluation of heating and cooling concepts, seeking to achieve a global optimum of interior thermal and humidity comfort, useful cooling-energy use, and the building's total primary-energy use for heating, cooling, ventilation, and lighting. Under this premise, ambitious planning concepts stand at the crossroads of economic sustainability, legislative restrictions, rising energy costs, the shortage of primary energy sources, and the demand for a high comfort of use. An integral step-by-step approach towards low-energy cooling concepts supports the planner in order to find the best solutions for a specific project. With careful system matching in the first project phase, individual solutions can be relatively easily implemented and the overall concept remains technically manageable.

Sustainable and environmentally responsible nonresidential building concepts:

- Guarantee enhanced visual, acoustic, and thermal comfort and therefore provide a high-quality workplace environment, which improves the occupant's productivity and reduces the impact of the built environment on his/her health.
- Harness the building's architecture and physics, in order to considerably reduce the annual heating and cooling demand (building envelope, day-lighting concept, natural ventilation, passive heating, and cooling technologies).
- Put emphasis on a highly energy-efficient heating and cooling plant with a significantly reduced auxiliary energy use for the generation, distribution, and delivery of heating and cooling energy. The applied components and technologies are soundly orchestrated by optimized operation and control strategies.
- Use less valuable primary energy, e.g., more renewable energy from environmental heat sources and sinks, solar power, biomass, etc.

D. E. Kalz and J. Pfafferott, *Thermal Comfort and Energy-Efficient Cooling*
of Nonresidential Buildings, SpringerBriefs in Applied Sciences and Technology,
DOI: 10.1007/978-3-319-04582-5_8, © The Author(s) 2014

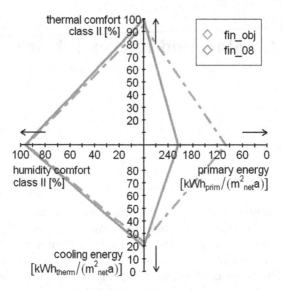

Fig. 8.1 Building signature. This building signature shows results from the Finnish monitoring campaign and its evaluation in accordance with the guidelines given in this guidebook. The thermal indoor environment meets the requirements of class II. The useful cooling energy meets the building-physical requirements on summer heat protection. Only the primary-energy demand of the building is higher than the target value and does not meet the requirements. Measurements (*solid line*), objective (*dotted line*)

8.1 Success Factors for Low-Energy Cooling Concepts

Under this premise, a holistic approach is proposed for the evaluation of heating and cooling concepts, seeking to achieve a global optimum of (1) interior thermal comfort, (2) interior humidity comfort, (3) useful cooling-energy use, and (4) the building's total primary-energy use for heating, cooling, ventilation, and lighting.

Figure 8.1 illustrates an individual building signature correlating cooling-energy use ($kWh_{therm}/(m^2_{net}a)$), the building's total primary-energy use for heating, cooling, ventilation, and lighting (kWh_{prim}/m^2a), and thermal and humidity comfort classifications in accordance with EN 15251:2007-08. The green triangle represents the target objective for these three parameters and the arrows indicate the direction of the optimum.

Occupant Thermal and Humidity Comfort. Occupant thermal comfort assessments of the buildings in summer are evaluated in accordance with the European EN 15251:2007-08 guideline. The building signatures present the time at the required comfort class during occupancy. Thermal comfort is evaluated with the proposed methodology in accordance with the

- Adaptive-comfort approach for building concepts with passive cooling and the
- PMV-comfort approach for building concepts with water-based mechanical and mixed-mode cooling.

The target objective for the comfort class is defined during the design stage of the building, i.e., class III for the building in Greece; class II for the buildings in Germany, France, the Czech Republic, Romania; and class I for the buildings in Italy, Finland, and Denmark. Then, thermal comfort measurements are evaluated correspondingly. The comfort class is guaranteed if recorded temperature values remain within the required comfort class during 95 % of the occupancy time.

Cooling-Energy Use. Measurements of useful cooling energy are derived from the long-term monitoring campaigns—carried out by the particular ThermCo partners. If measurements are not available, simulation results or calculations are presented. Cooling-energy use depends on the building's architecture, user behavior, climate, and the potential of the heat sink employed. Therefore, the cooling load (W/m^2) increases from north to south mainly due to higher temperatures and—to a lesser extent—due to higher solar heat gains. Consequently, the target objectives for the cooling-energy use vary due to climate and building concepts. For the building assessment, objective cooling-energy values are taken from the simulation study in Chap. 7, representing a typical low-energy nonresidential building.

Primary-Energy Use. The primary-energy consumption of the buildings considers the heating and cooling plant as well as ventilation and lighting—and was limited to a value of 100 $kWh_{prim}/(m_{net}^2 a)$. If not stated otherwise, plug loads are not included. The primary-energy approach allows for comparing concepts that use different energy sources such as fossil fuels, electricity, environmental energy, district heat, waste heat, and biomass. The primary-energy factors are 2.5 MWh_{prim}/MWh_{final} for electricity and 1.0 MWh_{prim}/MWh_{final} for fossil fuels.

Energy Efficiency. Almost all buildings investigated proved to be energy efficient. The total primary-energy consumption for heating, cooling, ventilation, and lighting ranges between 32 and 240 $kWh_{prim}/(m_{net}^2 a)$. The night-ventilation concept provides useful cooling energy in the range of 5–18 $kWh_{therm}/(m_{net}^2 a)$. If an earth-to-air heat exchanger is employed, the cooling energy is supplied with an energy efficiency of SPF 20 kWh_{therm}/kWh_{prim} (related to primary energy). The mechanical ventilation systems provide cooling energy with an efficiency of SPF 0.5–15 kWh_{therm}/kWh_{fin} (again related to primary energy). The environmental cooling systems provide useful cooling energy in the range of 5–44 $kWh_{therm}/(m_{net}^2 a)$, with an efficiency of SPF 1.3–8.0 kWh_{therm}/kWh_{fin} of the entire cooling system (related to final energy use).

Conclusion. In conclusion, a well-designed and well-operated building provides thermal and humidity comfort in compliance with the required comfort class of EN 15251:2007-08, with a reduced cooling-energy demand (below values derived from the simulation study) as well as an overall efficient HVAC and lighting concept, which results in a limited primary-energy use of 100 $kWh_{prim}/m^2 a$.

8.2 Integral Design of Low-Energy Cooling Concepts

Ambitious planning concepts stand at the crossroads of political conditions, rising energy costs, the shortage of primary energy sources, and the demand for a high comfort of use. At the same time, the new or renovation project shall be economically sustainable. The flood of information, a wide range of planning concepts and available technical building solutions, as well as personal preferences of stakeholders and varied expectations on the particular project make it difficult to decide for an optimal combination.

In Article 9 of the Directive 2010/31/EU on the "energy performance of buildings directive," the European Union calls on its member states to "ensure that by 31 December 2020, all new buildings are nearly zero-energy buildings." Following this requirement on new buildings, retrofit buildings are also getting increasingly energy efficient. This development is accompanied by low-energy cooling systems, which have been developed for more than 20 years.

Retrofit Projects. Processes, concepts, and technologies were evaluated in the study "Advances in Housing Retrofit" (Herkel and Kagerer 2011). Many technical solutions inform about the need for the careful planning and implementation of structural reorganization, in particular with regard to airtightness and thermal bridging. Successful examples for subsequent installations of ventilation systems show a great variety technical solutions with free ventilation, exhaust air only, and supply-and-exhaust ventilation with heat recovery. Based on a reduced heating and cooling load, innovative system concepts can be used very efficiently in order to meet energy demand. The applied air- and water-distribution systems are driven either by conventional plant technology or by small CHP units, heat pumps, biomass boilers, or combi-systems. Furthermore, all projects use solar-thermal or photovoltaic energy in combination. The energy-economic analysis of 60 international projects shows the high potential of a sophisticated building renovation with modified plant technology.

New Buildings. The study "Towards Zero-Energy Solar Buildings" (Voss and Musall 2011) presents international projects for carbon-neutral living and working. In principle, a building can be run energy-self-sufficient. The Monte Rosa Hut (2009) in the Valais Alps (Switzerland) and the Self-Sufficient Solar House (1992) in Freiburg (Germany) are prominent examples of this extreme form. However, in order to minimize the need for energy saving and, on the other hand, the risk of a supply disruption, zero-energy buildings (Sartori et al. 2012) are usually designed with net coupling.

The Role of Costs in Planning. A major obstacle in the implementation of energy-demanding projects are the costs. However, the positive experiences of many documented retrofit (EnOB 2013b) and new (EnOB 2013a) projects clearly indicate that they can be implemented economically. Unfortunately, the planning of projects in day-to-day practice is too often based on individual decisions without a vision for the entire project. Thus, it is often difficult to harmonize building and equipment and to find a cost-effective solution in an optimal way.

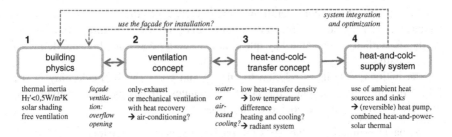

Fig. 8.2 Integral step-by-step approach toward low-energy concepts

A comprehensive analysis of the cost-effectiveness of individual measures in retrofit projects shows that many of them can be implemented beneficially, if measures for building renovation are pending anyway (BBR 2008). Accordingly, an insulation thickness between 12 and 36 cm is economically meaningful, depending on the use of the basement wall, exterior wall, or roof. Today, a renovation with triple glazing—wherever possible in construction—is state of the art and good business sense in any case. The use of ventilation systems for a good energy performance of buildings is becoming the standard. Whether or not a simple exhaust-air system or a supply-and-exhaust-air system with heat recovery is used is often a question of feasibility rather than costs.

A profitability calculation is part of the planning of new buildings, analyzing the overall costs for complete heating, ventilation, and air-conditioning concepts, and including the applied heat transfer systems. Often, the differences in annual costs are minimal in alternative concepts. This is finally reflected in the fact that differing concepts are applied to structures with similar demands. In addition to (subjective) preferences, investment costs—and not the annual energy costs—often turn the balances for a specific variant. Therefore, building and energy concepts do not perfectly match in many projects, and the energy efficiency of the overall solution might not correspond to the economic energy optimum either. The study "Costs and Potentials of Greenhouse Gas Abatement in Germany" (McKinsey 2009) clearly indicates that most innovative, energy-efficient building technologies are economically feasible today.

Hence, a comprehensive analysis of the total costs often leads to an economically energy-efficient solution. Therefore, the costs of individual measures should not be overestimated at the beginning of a project. In the preliminary design, energy-optimized scenarios are often developed, which later proves to be the most economical solution.

Integral Design Approach. As we have seen in Chaps. 6 and 7, a joint assessment of building physics and systems engineering can be achieved with a common representation of comfort (users), net-energy (building standard), and primary-energy demand (efficiency), separately for the heating and cooling cases.

How can building physics and innovative plant technology be matched in an optimal manner under this legal framework by using established standards and guidelines? Practical considerations often determine the energy concept in advance. A step-by-step approach may get to an optimal system integration. Based on the individual aspects and key questions of energy concepts, Fig. 8.2 defines a critical path.

The major decisions in relation to energy and costs can be made only in the initial planning stages, since the costs of changes are always higher with progressive provisions in the course of the project:

- The basic evaluation establishes the basis for objective decisions.
- The preliminary draft should present various system designs. The system can be selected by means of a decision matrix that considers its objectives.
- The principal components can be specified in the design phase.
- In the following planning, approval, construction, and commissioning phases, we look at the system tuning.

It is of crucial importance for the project's success that technical aspects are considered both individually and collectively: Significant interfaces arise between the issues of building envelope, ventilation design, delivery system, and energy provision. Though an optimal solution can be achieved across the trades, the first planning steps follow trade-by-trade.

The decision is to be made in the individual planning steps. The following indications can assist in developing a low-energy cooling concept:

- For practical construction considerations, H_T' values (mean heat-transfer value of building envelope) between 0.4 and 0.5 W/m^2 K can be achieved in a technically and economically easy manner.
- Often, effective sun-protection concepts can be integrated as part of the façade.
- Appropriate sun protection in connection with the construction standard described above and a (hybrid) ventilation concept may provide a comfortable indoor environment without additional cooling in residential buildings. In office buildings, the reduction of the (specific) cooling load is a key condition for the energy-efficient use of environmental energy for cooling.
- A mechanically assisted ventilation concept is a basic requirement for energy-efficient buildings in winter and can favorably be used for low-energy cooling in summer, too.
- Reduced heating and cooling loads enable the use of low-temperature radiant or ventilation systems very efficiently.
- If the ventilation (as adjunct) is also used for heating, cooling, or dehumidification, the interaction of water- and air-based handover systems must be taken into account. On the other hand, it is essential to ensure that the temperature level of the respective heating or cooling registers corresponds to the one in the water-based system.
- Mainly in retrofit but also in new buildings, it should be examined whether heating, cooling, and air distribution can be realized on the façade.

Under these conditions, many combinations that can be addressed only as examples are available for the provision of energy. Systems for low-energy cooling and heating are of particular interest:

- If night-ventilation for low-energy cooling is exclusively used, volume flow, pressure drops, and operation time have to be defined accurately in order to minimize the use of electrical fan power. These systems may be supplemented with evaporative cooling during daytime.
- Ground-source reversible heat pumps mainly use environmental energy for heating and cooling. They can therefore be operated economically, but are a relatively expensive investment. With the correct design of the overall system, the heating and cooling primary-energy production is much cheaper than conventional systems with boilers and simple compression chillers.
- A combined heat, cooling, and power generation is only economically viable if the energy-demand structure harmonizes very well with the system design. Whenever heat and cold have to be provided simultaneously, a suitable heat-pump system with the ability to shift the heat is often the better solution.

In the interests of sustainable energy concepts, focusing exclusively on the building standard is as unpromising as the exclusive focus on the most efficient energy supply. We may apply the 80/20 rule for both sets of measures: "building physics" and "plant engineering"—according to which 80 % of success is achieved with 20 % of expenses, while with the remaining 20 % can only be achieved with 80 % of the possible effort.

With careful system matching in the first project phase, individual solutions can be relatively easily implemented and the overall concept remains technically manageable. Hence, low-energy cooling concepts can be favorably implemented in low-energy buildings under market conditions.

References

BBR (2008) Bewertung energetischer anforderungen im lichte steigender energiepreise für die EnEV und die KfW-förderung. Bundesamt für Bauwesen und Raumordnung, Bonn

EnOB (2013a) Energieoptimiertes Bauen im Neubau. www.enob.info/de/neubau. Accessed Dec 2013

EnOB (2013b) Energieoptimiertes Bauen in der Sanierung. www.enob.info/de/sanierung. Accessed Dec 2013

Herkel S, Kagerer F (eds) (2011) Advances in housing retrofit. Report on IEA task 37 advanced housing renovation with solar and conservation, Fraunhofer ISE, Freiburg

McKinsey & Company (2009) Kosten und Potenziale der Vermeidung von Treibhausgasemissionen in Deutschland. Aktualisierte Energieszenarien und -sensitivitäten, McKinsey & Company

Sartori I, Napolitano A, Voss K (2012) Net zero energy buildings: A consistent definition framework. Energ Buildings 48:220–232

Voss K, Musall E (eds) (2011) NULLENERGIEGEBÄUDE. Report on IEA task 40 toward zero-energy solar buildings, DETAIL Green Books, München

Glossary

Auxiliary Energy Auxiliary energy is necessary to harvest the heating/cooling energy from the environmental heat source/sink (primary hydraulic circuit), as well as to distribute the energy through the building and, finally, to deliver the heating/cooling energy to the offices and rooms via thermo-active building systems. In the primary hydraulic circuit, auxiliary energy uses accounts for the submerged pump (groundwater well), for the brine pump (borehole heat exchangers and energy piles), as well as for the circulation pump, and for the fan of the wet-cooling tower. In the secondary hydraulic circuit, auxiliary energy is used to operate the distributor, the thermal storage loading, and the circulation pumps.

COP The ratio of the power output to the power input of a system. Also, see "SPF."

Energy Consumption The actually measured quantity of energy needed for heating, cooling, ventilation, hot water heating, lighting, appliances, etc.

Energy Demand Calculated quantity of energy for all applications and given end use. Energy to be delivered by an ideal energy system (no system losses are taken into account) in order to provide the required service to the end user, e.g., to maintain the required internal set-point temperature of a heated or cooled space.

Exergy Energy consists of exergy and anergy. Exergy is the part of the energy that can be transformed into any form of energy within defined boundary conditions. Anergy is the part of energy that cannot be transformed into exergy.

Final Energy Energy that is delivered to the building (fossil fuel, electricity, etc.) from the last market agent.

Heated Net Floor Area Specific primary and final energy use is related to the *heated net floor area* of buildings: the sum of all heated areas within the building, including heated corridors and internal stairways but not unheated rooms, in Germany, in accordance with [DIN 277-1:2005-01].

D. E. Kalz and J. Pfafferott, *Thermal Comfort and Energy-Efficient Cooling of Nonresidential Buildings*, SpringerBriefs in Applied Sciences and Technology, DOI: 10.1007/978-3-319-04582-5, © The Author(s) 2014

Low-Energy Building Buildings with the explicit purpose to use less energy than standard buildings.

Low-Exergy "Low-exergy (LowEx) systems" are defined as heating or cooling systems that allow the use of low-valued energy as their energy source. In practice, this means systems that provide heating or cooling energy close to room temperature with low heat-flow density.

Operative Room Temperature This is the arithmetic mean of dry-bulb and surface temperature of a room if air velocity is lower than 0.2 m/s.

Primary Energy Energy that has not been subjected to any conversion or transformation process. These factors vary for each country. The primary energy conversion factors for this study were selected to be 2.5 for electricity, 0.2 for biomass, 1.1 for fossil fuels, and 0.7 for district heat from cogeneration, in accordance with [DIN V 18599-1:2007-02]. The primary energy conversion factor for electricity is subjected to modifications in accordance with the development of each national electricity market.

Residential Sector Public and private community accommodation, i.e., private housing, flats, student accommodation, etc.

Service Sector The service sector, also referred to as the tertiary sector, includes the public sector as well as the non-industrial/manufacturing (private) sectors such as public administration, education and health, banking and finance, and trade. In the context of this book: public buildings (e.g., health care, education, administration) and commercial buildings (e.g. retail, office, hotel, leisure).

SPF The heating and cooling systems are evaluated in terms of energy efficiency, according to the defined balance boundaries. Efficiency is described by the coefficient of performance (COP_h) in the heating mode and by the energy-efficiency ratio (EER) in the cooling mode (in Europe COP_c). The COP is the ratio of the useful energy acquired, divided by the energy applied, such as auxiliary electricity needed for the pumps or for the compressor of the heat pump. The approximated COP for the heating and cooling seasons, respectively, is described by the *seasonal performance factor (SPF)*, in accordance with [ASHRAE Handbook 2000] and [DIN 18599-1:2007-02], taking the system operation and part-load impacts into account. See also "COP."

TABS Thermo-active building systems are construction elements thermally activated by water or airborne systems that operate with small temperature differences between room air and the thermally activated building component, allowing the use of low-temperature heat sources and sinks.

Thermal Cooling Energy Thermal energy necessary to cover the load in order to achieve a certain room temperature.

Useful Energy Portion of final energy that is actually available to the consumer for respective use after final conversion, e.g., "thermal cooling energy."